LEBESGUE'S THEORY *of* INTEGRATION

LEBESGUE'S
THEORY
of
INTEGRATION

*Its Origins
and Development*

THOMAS HAWKINS

Chelsea Publishing Company
THE BRONX, NEW YORK, N.Y.

53188

Second Edition

Copyright © 1970 by the Regents of the University of Wisconsin
Copyright © 1975 by the Regents of the University of Wisconsin
Second edition printed 1975 on special long-life acid-free paper

Library of Congress Cataloging in Publication Data

Hawkins, Tomas, 1938-
   Lebesgue's theory of integration.

   Bibliography: p.
   1.   Integrals, Generalized.   I.   Title.

QA312.H34   1975   515'.43   74-8402
ISBN   0-8284-0282-5

Printed in the United States of America

Preface

THIS BOOK grew out of my dual interest in the mathematics of integration theory and in the nature of its historical evolution. It is well known that modern theories of measure and integration have their roots in the work of the French mathematician Henri Lebesgue. Although a relatively small amount of space is devoted to actually discussing Lebesgue's work, it represents the focal point of the entire book. My objective has been to place Lebesgue's early work on integration theory (1901–10) within its proper historical context by relating it, on the one hand, to the developments during the nineteenth century that motivated it and gave it significance and, on the other hand, to the contributions made in this field by his contemporaries. It is my hope that this volume will engender an appreciation for Lebesgue's genius balanced by an appreciation for the manner in which his work builds upon that of many other mathematicians.

A few remarks are in order concerning the format. The mathematical notation employed has been chosen with the intention of causing a minimum of distraction to the flow of ideas. Thus I have followed or approximated an author's notation only when this was convenient. Even in cases of notational discrepancy, however, the reader should experience little or no difficulty relating the discussion to the original papers. The glossary contains technical terms as well as frequently used expressions peculiar to this book, e.g. "the Fundamental Theorem II" and "Riemann's integrability condition (R_1)."

Finally, I wish to thank Professors R. Creighton Buck and Erwin Hiebert of the University of Wisconsin for their encouragement and advice, my wife, Marie, for her invaluable secretarial assistance, and the National Science Foundation for financial support to carry out the research that made the following study possible. I am also indebted to Professor Gerald S. Goodman of the University of Florence for calling my attention to the consequential error on page 104 of the first edition.

Contents

Introduction   ix

Chapter 1   *Riemann's Theory of Integration*   3

    1.1 THE EIGHTEENTH-CENTURY CONCEPTION OF A
    FUNCTION AND TRIGONOMETRIC SERIES REPRE-
    SENTATIONS, 3
    1.2 CAUCHY'S CONCEPTION OF THE INTEGRAL, 9
    1.3 DIRICHLET'S CONTRIBUTION, 12
    1.4 RIEMANN, 16

Chapter 2   *The Development of Riemann's Ideas: 1870–80*   21

    2.1 TRIGONOMETRIC SERIES AND TERM-BY-TERM IN-
    TEGRATION, 21
    2.2 THE REFORMULATION OF RIEMANN'S INTEGRA-
    BILITY CRITERIA, 28
    2.3 DISCUSSIONS OF THE DIFFERENTIABILITY PROP-
    ERTIES OF CONTINUOUS FUNCTIONS, 42

Chapter 3   *Set Theory and the Theory of Integration*   55

    3.1 NOWHERE DENSE SETS AND THEIR MEASURE-
    THEORETIC PROPERTIES, 55
    3.2 THE INTRODUCTION OF OUTER CONTENT, 61
    3.3 CANTOR'S DEVELOPMENT OF THE THEORY OF SETS
    AND ITS APPLICATION TO THE THEORY OF INTE-
    GRATION, 71
    3.4 CURVE LENGTH AND THE INTEGRAL, 79

Chapter 4   *The End of the Century: A Period of Transition*   86

    4.1 THE INTRODUCTION OF THE CONCEPT OF MEASURABILITY, 86
    4.2 BOREL'S THEORY OF MEASURE, 97
    4.3 SCHOENFLIES' REPORT ON THE THEORY OF SETS, 106
    4.4 RESEARCHES ON TERM-BY-TERM INTEGRATION OF NONUNIFORMLY CONVERGING SERIES, 110

Chapter 5   *The Creation of Modern Integration Theory*   120

    5.1 LEBESGUE'S THEORY OF INTEGRATION, 120
    5.2 THE WORK OF VITALI AND YOUNG ON MEASURE AND INTEGRATION, 146
    5.3 FUBINI'S THEOREM, 154

Chapter 6   *Pioneering Applications of the Lebesgue Integral*   163

Epilogue   *The Lebesgue-Stieltjes Integral*   179

Appendix   *Dini's Theorem on the Differentiability of Continuous Functions*   197

    Glossary   200

    Special Symbols   204

    List of Abbreviations   205

    Bibliography   208

    Index   225

Introduction

LEBESGUE MAY be said to have created the first genuine theory of integration. Various definitions, theorems, and examples existed prior to his work, but they lacked the coherence and completeness of a true theory. Nevertheless, these pre-Lebesgue contributions paved the way for a sophisticated theory of integration. Specifically, they provided Lebesgue with both (1) a fully developed measure-theoretic point of view and (2) a number of theoretical "problems" that had been discovered within the context of Riemann's definition of the integral (although they were not seriously regarded as problematic at the time). To appreciate what Lebesgue did and why he did it, his work must be regarded within the context of the historical developments that produced (1) and (2). For this reason, a substantial portion of the book is devoted to tracing these developments. It would be difficult to treat separately the developments that produced (1) and (2) without distorting the historical perspective, since they overlap considerably, and this approach was therefore not attempted. But it is worthwhile to begin with an overview of the entire book which clarifies the significance of (1) and (2) separately.

The historical significance of the development of a measure-theoretic viewpoint is that it provided new ways of looking at the Cauchy-Riemann definition of the definite integral—ways that made its generalizability more apparent. Riemann's theory of integration (1854) was derived from Cauchy's by weakening as far as possible the assumptions on the function to be integrated. That is, whereas Cauchy restricted himself to continuous functions, Riemann defined a bounded function $f(x)$ to be integrable on $[a,b]$ if and only if the Cauchy sums

$$S = \sum_{k=1}^{n} f(t_k)(x_k - x_{k-1}),$$

where $a = x_0 < x_1 < \cdots < x_n = b$ and $t_k \in [x_{k-1}, x_k]$, approach a unique limiting value as the norm of the partition approaches 0. This unique limiting value is then by definition $\int_a^b f(x)dx$.

Although what Riemann did may now appear to be rather straightforward and almost trivial, historically it represented a bold, perceptive departure from the past because it involved a radically different conception of a function. At the time Riemann's theory thus appeared as the ultimate in generality: his integrability condition was the weakest under which the traditional definition, based on Cauchy sums, retains a meaning; the integrability condition was so weak that the integral concept could be extended to functions whose points of discontinuity form a dense set—functions whose very existence had not even been contemplated by most mathematicians. A further generalization therefore seemed unthinkable. Unthinkable, that is, as long as Cauchy sums were regarded as the only approach to the definition of the integral. It is in this respect that measure-theoretic ideas became important, for they provided a new basis for defining the Riemann integral.

Let us first recall some definitions. If S denotes a bounded set of real numbers, the *outer content* of S is the real number $c_e(S)$ defined as follows. Let $I_1, I_2, \cdots, I_n$ denote a finite set of intervals which cover S in the sense that $S \subset \bigcup_{k=1}^{n} I_k$. Then $c_e(S)$ denotes the greatest lower bound of all real numbers of the form $\sum_{k=1}^{n} L(I_k)$ where the I_k cover S and $L(I_k)$ is the length of I_k. Similarly, the *inner content* of S, $c_i(S)$, is defined as the least upper bound of $\sum_{k=1}^{n} L(I_k)$, where $I_k \cap I_{k'} = \varnothing$ and $\bigcup_{k=1}^{n} I_k \subset S$. A set is said to be *Jordan-measurable* if $c_i(S) = c_e(S)$, and the *content* of a measurable set is $c(S) = c_i(S) = c_e(S)$. When it is clear that we are referring to content rather than to Lebesgue measure, we shall drop the "Jordan" and say simply "measurable." It should be noted that these definitions extend naturally to higher dimensions.

These notions, which had been introduced by the early 1890's, suggested two new characterizations of Riemann's integrability condition. The first is based on the "geometrical" view of the integral of a function in terms of the area bounded by its graph. For f defined and bounded on the interval $[a,b]$, let E denote the set of points in the plane bounded by the graph of f, the lines $x=a$ and $x=b$, and the x-axis. Then f is Riemann-integrable if and only if the set E is Jordan-measurable, and

$$(1) \qquad \int_a^b |f| = c(E), \qquad \int_a^b f = c(E^+) - c(E^-),$$

where E^+ and E^- denote the parts of E above and below the x-axis, respectively.

The second characterization of Riemann's integrability condition derives from the fact that the lower and upper integrals of f—that is,

$$\underline{\int}_a^b f \text{ and } \overline{\int}_a^b f \text{—are required to be equal. Now } \underline{\int}_a^b f \text{ and } \overline{\int}_a^b f \text{ rep-}$$

resent the least upper bound of the numbers L and the greatest lower bound of the numbers U, respectively, for

$$L = \sum_{i=1}^n m_i(x_i - x_{i-1}) \quad \text{and} \quad U = \sum_{i=1}^n M_i(x_i - x_{i-1}),$$

where $a=x_0<x_1< \cdots <x_n=b$ denotes a partition of $[a,b]$, and m_i and M_i denote, respectively, the greatest lower bound and the least upper bound of $f(x)$ on $[x_{i-1},x_i]$. The introduction of the concept of a measurable set brought about the following variation in this characterization of integrability: Consider the more general sums

$$(2) \qquad L = \sum_{i=1}^n m_i c(E_i) \quad \text{and} \quad U = \sum_{i=1}^n M_i c(E_i),$$

where the E_i are measurable, pairwise disjoint sets such that $[a,b]= \bigcup_{i=1}^n E_i$. Then the least upper bound of L and the greatest lower bound of U are still the lower and upper integrals of f, and so Riemann's integrability condition can be stated in terms of these more general sums.

Both of these characterizations made it possible to see that a generalization of the concepts of measure and measurability would afford a generalization of the concepts of the integral and integrability. Suppose, in other words, that $\mathfrak{M}$ denotes a class of measurable sets which contains the class of Jordan-measurable sets, and suppose that a measure, $m(E)$, has been defined for all members of $\mathfrak{M}$ in such a manner that $m(E)$ coincides with $c(E)$ when E is Jordan-measurable. Then the first characterization of Riemann-integrable functions, equation (1), suggests that the concept of integrability could be extended to any bounded function f whose corresponding set E belongs to $\mathfrak{M}$. The integral of f would then be defined by $\int_a^b f = m(E^+) - m(E^-)$. Similarly, the second characterization, equation (2), suggests defining lower and upper integrals $*\underline{\int}_a^b f$ and $*\overline{\int}_a^b f$ with respect to $\mathfrak{M}$ as the least upper bound of L^* and the greatest lower bound of U^* for the sums

$$L^* = \sum_{i=1}^n m_i m(E_i) \quad \text{and} \quad U^* = \sum_{i=1}^n M_i m(E_i),$$

where now the E_i belong to $\mathfrak{M}$. Then

$$\underline{\int}_a^b f \leq *\underline{\int}_a^b f \leq *\overline{\int}_a^b f \leq \overline{\int}_a^b f$$

and f could be defined to be integrable provided $*\underline{\int}_a^b f = *\overline{\int}_a^b f$.

Clearly, both of these definitions, based on a broader concept of measurability, represent generalizations of the Riemann integral; when $\mathfrak{M}$ denotes the class of Lebesgue-measurable sets, the definitions yield the Lebesgue integral for bounded functions.

It was essentially through such considerations that Lebesgue (and, independently, W. H. Young) obtained their generalizations of the integral—after Émile Borel suggested the sort of properties a generalized measure ought to have. Indeed, once these ideas had been developed, it was rather inevitable that someone would eventually apply them to the integral concept, and it is for this reason that special attention is paid to the development of measure-theoretic notions before Lebesgue and their relation to the theory of integration.

For Lebesgue, however, the generalized definition of the integral represents only the beginning and the least profound part of his contribution to integration theory. What made the initial discovery important was that he was able to recognize in it an analytical tool capable of dealing with—and to a large extent overcoming—the unresolved problems that had arisen in connection with the old theory of integration. As Lebesgue once explained, "a generalization made not for the vain pleasure of generalizing but in order to solve previously existing problems is always a fruitful generalization" [1966: 194]. Indeed, the unresolved problems motivated all of Lebesgue's major results.

The first such problem was raised unwittingly by Fourier in 1822: (A) If a bounded function f can be represented by a trigonometric series, is that series the Fourier series of f? Closely related to (A) is: (B) When is the term-by-term integration of an infinite series of functions permissible? That is, when is it true that

$$\int_a^b \left(\sum_{n=1}^{\infty} u_n(x) \right) dx = \sum_{n=1}^{\infty} \int_a^b u_n(x) dx?$$

Fourier had assumed the answer to (B) is Always and had used (B) to prove that the answer to (A) is Yes. By the end of the nineteenth century it was recognized that (B) is not always true even for uniformly bounded series precisely because $f(x) = \sum_{n=1}^{\infty} u_n(x)$ need not be Riemann-integrable; furthermore, the positive results that were achieved required extremely long proofs. These developments, however, paved the way for Lebesgue's elegant proof that (B) is true for any uniformly bounded series of Lebesgue-integrable functions $u_n(x)$. And, by applying this result to (A), Lebesgue was able to affirm Fourier's belief that the answer is Always.

Another source of difficulties was what we have termed the Fundamental Theorem II, viz., the assertion that $\int_a^b f'(x) dx = f(b) - f(a)$. The work of Ulisse Dini and Vito Volterra made it clear that functions exist which have bounded, nonintegrable derivatives, so the Fundamental Theorem II becomes meaningless for these functions. Later further classes of functions with this property were discovered. Additional problems arose in connection with Axel Harnack's extension of the Riemann integral to unbounded functions because continuous monotonic functions with densely distributed intervals of invariability were discovered. These functions provided examples of Harnack-integrable derivatives for which the Fundamental Theorem II is false. Some theorems due to Lebesgue largely resolved these problems.

The existence of the above-mentioned class of monotonic functions also raised the question: When is a continuous function an integral? This prompted Axel Harnack to introduce the property that has since been termed "absolute continuity"; and during the 1890's absolute continuity came to be regarded as the characteristic property of absolutely convergent integrals, although no one was actually able to show that every absolutely continuous function is an integral. Lebesgue, however, perceived that this is precisely the case when integrals are taken in his sense.

A deeper familiarity with infinite sets of points had led to the discovery of the problems connected with the Fundamental Theorem II.

The nascent theory of infinite sets also stimulated an interest in the meaningfulness of the classical formula $L = \int_a^b [1+(f')^2]^{1/2}$ for the length of the curve $y = f(x)$. Paul du Bois-Reymond, who initiated an interest in the problem, was convinced that the theory of integration is indispensable for the treatment of the concepts of rectifiability and curve length within the general context of the modern function concept. But by the end of the nineteenth century this view appeared untenable, particularly because of the criticism and counterexamples given by Ludwig Scheeffer. Lebesgue was quite interested in this matter and was able to use the methods and results of his theory of integration to reinstate the credibility of du Bois-Reymond's assertion that the concepts of curve length and integral are closely related.

Lebesgue's work on the Fundamental Theorem II and on the theory of curve rectification played an important role in his discovery that a continuous function of bounded variation possesses a finite derivative except possibly on a set of Lebesgue measure zero. This theorem gains in significance when viewed against the background of the century-long discussion of the differentiability properties of continuous functions. During roughly the first half of the nineteenth century, it was generally thought that continuous functions are differentiable at "most" points, although continuous functions were frequently assumed to be "piecewise" monotonic. (Thus, differentiability and monotonicity were linked together, albeit tenuously.) By the end of the century this view was discredited, and no less a mathematician than Weierstrass felt there must exist continuous monotonic functions that are nowhere differentiable. Thus, in a sense, Lebesgue's theorem substantiated the intuitions of an earlier generation of mathematicians.

Riemann's extension of the integral concept also raised problems in connection with the classical theorem positing the identity of double integrals and iterated integrals, i.e.,

$$\int_R f(x,y)dR = \int_a^b \left[\int_c^d f(x,y)dx \right] dx = \int_c^d \left[\int_a^b f(x,y)dx \right] dy,$$

where R is the rectangle determined by $a \le x \le b$ and $c \le y \le d$. It was soon discovered that when $f(x,y)$ is integrable over R, the functions $x \to f(x,y)$ and $y \to f(x,y)$ can fail to be integrable on dense sets of y and x, respectively—sets with positive outer content. The classical formulation therefore had to be modified. And it was later discovered that the modifications become even more drastic when unbounded functions are considered. Although Lebesgue did not resolve this infelicity, it was his treatment of the problem that formed the foundation for Fubini's well known theorem that did restore in essence the simplicity of the classical formulation.

By the end of the first decade of the twentieth century, Lebesgue's ideas had begun to catch on, and the number of mathematicians engaged in work relating to the Lebesgue integral started to increase rapidly. Lebesgue's own work during this decade—particularly his applications of the new integral to trigonometric series—was the chief reason, but the pioneering researches of other mathematicians—Guido Fubini, Pierre Fatou, Ernst Fischer, and F. Riesz—also contributed substantially to the trend. The early applications of the Lebesgue integral are discussed and related in Chapter 6.

In the final section of the book, the developments leading up to the creation of the so-called Lebesgue-Stieltjes integral by J. Radon in 1913 are traced. This is an appropriate epilogue because Lebesgue's early work (1901–10) played an important role in bringing about its creation and because the fusion of the integrals of Lebesgue and Stieltjes, based as it is upon the concept of a countably additive set function, helped lay the foundations for modern abstract theories of measure and integration. Radon's paper represented a real triumph for Lebesgue's ideas, for it indicated that they remain viable in a much more general setting.

LEBESGUE'S THEORY *of* INTEGRATION

1

Riemann's Theory of Integration

The theory of integration proposed by Bernhard Riemann in 1854 was based upon and motivated by his acceptance of the modern concept of a function as any correspondence $x \rightarrow f(x)$ between real numbers. In this chapter we shall briefly review the developments which produced this concept[1] and their connection with the theory of integration. Special attention is paid to the changing meaning attached to the term "discontinuous function" as a consequence of these developments. This change in meaning is particularly important because the history of integration theory after Cauchy is essentially a history of attempts to extend the integral concept to as many discontinuous functions as possible; such attempts could become meaningful only after the existence of highly discontinuous functions was recognized and taken seriously.

1.1 THE EIGHTEENTH-CENTURY CONCEPTION OF A FUNCTION AND TRIGONOMETRIC SERIES REPRESENTATIONS

Although the notion of a function did not originate with Euler, it was he who first gave it prominence by treating the calculus as a formal theory of functions. In his *Introductio in analysin infinitorum* of 1748 he defined a function of a variable quantity as "an analytical

1. For further details and references, see [Jourdain 1906; 1913].

3

expression" composed in any way of that variable and constants [1748: 4]. The key to this definition is the notion of an analytical expression, which Euler evidently understood to be the common characteristic of all known functions. It was also Euler, however, who initiated a viewpoint that eventually led to the introduction of the modern concept of a function. In his pioneering study of partial differential equations of 1734, Euler admitted "arbitrary functions" into the integral solutions. And, in answer to Jean d'Alembert—who maintained that these arbitrary functions must be given by a single algebraic or transcendental equation in order to be the proper object of mathematical analysis—Euler clarified his earlier pronouncement by contending that the curves which the arbitrary functions represent need not be subject to any law but may be "irregular" and "discontinuous," i.e., formed from the parts of many curves or traced freehand in the plane.

It is important to observe that the term "discontinuous" as used by Euler and his contemporaries refers to a discontinuity in the analytical form of expression of the functional relationship: A function can be continuous in the modern sense and "discontinuous" in the sense of Euler. On the other hand, the possibility of arbitrary functions which are discontinuous in the modern sense at more than a finite number of points in a finite interval does not appear to have been seriously considered by anyone at this time. Attention was focused upon the fact that arbitrary functions are not determined by a single equation rather than upon their properties as correspondences $x \rightarrow f(x)$ between real numbers.

Central to the Euler-d'Alembert controversy was the question of the nature of the functions to be admitted as solutions to the partial differential equation of the vibrating string problem. The object was to determine the equation $y = F(x,t)$ of the string at time t. D'Alembert showed in 1747 that the solution must have the form

$$F(x, t) = \tfrac{1}{2}[f(x + t) + f(x - t)].$$

For $t = 0$, the initial shape of the string is given by the graph of $y = f(x)$. D'Alembert insisted that f must be a "continuous" function, i.e., given by a single equation, while Euler argued that this restriction was unnecessary and that f could be "discontinuous." A new dimension was introduced into the controversy by Daniel Bernoulli. On the basis of the physical principle that a vibrating string emits a fundamental frequency and its harmonic overtones, he concluded that the function f must be expressible as a series of the form

$$f(x) = a_1 \sin(\pi x/L) + a_2 \sin(2\pi x/L) + a_3 \sin(3\pi x/L) + \cdots,$$

where L denotes the length of the string. From a purely mathematical standpoint, Bernoulli's contention is to the effect that every arbitrary function can be represented for an entire interval of values of x by a trigonometric series; this contention was flatly rejected by most mathematicians, including Euler, d'Alembert, and Lagrange.

Interest in the notion of an arbitrary function was renewed early in the nineteenth century when Joseph Fourier (1768–1830) presented a paper on heat conduction to the Paris Academy of Science in 1807 and reaffirmed Bernoulli's contention. Fourier's paper was not published, and his ideas on trigonometric series representations first appeared in print in his *Théorie analytique de la chaleur* [1822]. In its most general form, Fourier's proposition can be stated in the following manner: Any (bounded) function f defined on $(-a,a)$ can be expressed in the form

$$f(x) = \tfrac{1}{2}\,a_0 + \sum_{n=1}^{\infty} [a_n \cos(n\pi x/a) + b_n \sin(n\pi x/a)],$$

where the coefficients are given by

$$a_n = (1/a) \int_{-a}^{a} f(x)\,\cos(n\pi x/a)dx,$$

$$b_n = (1/a) \int_{-a}^{a} f(x)\,\sin(n\pi x/a)dx.$$

Fourier was fully aware of the significance of his contention in the light of eighteenth-century discussions when he wrote:

> It is remarkable that we can express by convergent series . . . the ordinates of lines and surfaces which are not subject to a continuous law. We see by this that we must admit into analysis functions which have equal values . . . between two limits, even though on substituting in these two functions a number included in another interval, the results of the two substitutions are not the same. The functions which enjoy this property are represented by different lines which coincide in a definite portion only of their course. [1822: 199.]

An "arbitrary function" for Fourier was "a function completely arbitrary, that is to say a succession of given values, subject or not to a common law, and answering to all the values of x . . ." [1822: 432]. Furthermore, "the function $f(x)$ represents a succession of values or ordinates each of which is arbitrary . . . . They succeed each other in any manner whatever, and each of them is given as if it were a single quantity" [1822: 430].

These passages, particularly the last sentence above, may suggest that Fourier held an extremely general conception of a function. Actually this is not the case: Fourier's conception of a function was that of his eighteenth-century predecessors. His terminology hints at this, and in particular he used the term "discontinuous function" in the eighteenth-century sense. Thus, for example [1822: 340], he referred to the function that takes the value e^{-x} when x is nonnegative and the value e^x when x is negative as a discontinuous function, although it is continuous in the modern sense. As with Euler and his contemporaries, "arbitrary" was used by Fourier with reference to the formal expression of a function: an arbitrary function is one not given by a single equation. Such a function is not necessarily arbitrary in the sense of being possibly highly discontinuous or nondifferentiable. Indeed, Fourier seemed to think arbitrary functions are very well behaved. For example, he claimed that an arbitrary function $f(x)$ could be represented in the form

$$f(x) = (1/2\pi) \int_a^b f(\alpha)d\alpha \int_{-\infty}^\infty \cos(px - p\alpha)dp.$$

The formula itself is literally meaningless, since the improper integral fails to converge. But Fourier overlooked this point, and since x appears on the right-hand side only in $\cos(px-p\alpha)$, which represents a differentiable function, Fourier concluded that the function f "acquires in a manner by this transformation, all the properties of trigonometrical quantities; differentiations, integrations and summations of series thus apply to functions in general in the same manner as to exponential trigonometric functions" [1822: 434].

What, then, did the term "function" signify in Fourier's analysis? Certainly his own examples[2] indicate that for a *finite* number of values of x in a finite interval, a function represented by a trigonometric series might become discontinuous (in the modern sense) or might not be differentiable. It would seem, therefore, that Fourier had in mind functions whose graphs are smooth curves except for possibly a finite number of exceptional points within a finite interval.

Unlike Bernoulli, Fourier did not base the validity of his proposition about trigonometric series representation upon a physical principle. He realized the need to justify the proposition mathematically and pro-

2. See, for example, [Fourier 1822: 143–44, 150, 192, 194]; also, the function $e^{-|x|}$ given here.

vided two arguments. Both are based upon the idea that, for f an arbitrary function defined for x between $-\pi$ and π, if the equation

(1)
$$f(x) = \tfrac{1}{2} a_0 + \sum_{n=1}^{\infty} (a_n \cos nx + b_n \sin nx)$$

can be solved—i.e., if the values of the infinite number of unknowns $a_0, a_1, \cdots, b_1, b_2, \cdots$ can be determined—then the representation must be valid [1822: 198]. In the first proof [1822: 168–84] Fourier had to assume that $f(x)$ could be expanded in a power series in x; and then by nonrigorous but suggestive manipulations involving an infinite number of equations in an infinite number of unknowns, he determined the coefficients. For the second proof, however, he claimed complete generality [1822: 184–208]. The idea behind the proof had already been used by Lagrange for finite trigonometric series. To determine, for example, a_m, both sides of equation (1) are first multiplied by $\cos mx$. Both sides of the resulting equation are then integrated between $-\pi$ and π, with $\displaystyle\int_{-\pi}^{\pi} \sum_{n=1}^{\infty}$ replaced by $\displaystyle\sum_{n=1}^{\infty} \int_{-\pi}^{\pi}$ on the right-hand side— that is, term-by-term integration is assumed permissible. Then a_m is represented as the integral $\displaystyle(1/\pi) \int_{-\pi}^{\pi} f(x) \cos mx \, dx$.

Fourier's demonstration thus predicates the existence of the representation (1) for any (presumably bounded) function, a hypothesis which turns out to be invalid when the function concept is permitted complete generality in accordance with the modern viewpoint. But, if the validity of (1) is assumed, two further assumptions are still involved: (a) that the function f in (1) and its products with $\cos nx$ and $\sin nx$ possess integrals; and (b) that term-by-term integration of (1), or of the series resulting from multiplying (1) by $\cos nx$ or $\sin nx$, is permissible. It was not until much later that the general validity of term-by-term integration was seriously questioned (see 2.1).

But Fourier did recognize the need to justify the meaningfulness of representing the coefficients a_n and b_n as integrals, that is, the meaningfulness of the definite integral of an arbitrary function. During the eighteenth century, integration was conceived primarily as the inverse of differentiation rather than as the limit of a sum, although both viewpoints were recognized. The process of integrating a function f amounted to finding a primitive function, that is, a function F such that $F'=f$. The definite integral $\displaystyle\int_a^b f$ is then simply $F(b)-F(a)$. The existence of F is far from obvious, however, when f is arbitrary and not given by an equation. Probably with this in mind, Fourier returned to the sum-conception of the integral as an area:

> the coefficients . . . are the values of definite integrals . . . . This remark is important, because it shows how even entirely arbitrary functions may be developed in series of sines of multiple arcs. In fact, if the function $\phi(x)$ be represented by the variable ordinate of any curve whatever whose abscissa extends from $x = 0$ to $x = \pi$, and if on the same part of the axis the known trigonometric curve, whose ordinate is $y = \sin x$, be constructed, it is easy to represent the value of any integral term. We must suppose that for each abscissa x, to which corresponds one value of $\phi(x)$, and one value of $\sin x$, we multiply the latter value by the first, and at the same point of the axis raise an ordinate equal to the product $\phi(x) \sin x$. By this continuous operation a third curve is formed, whose ordinates are those of the trigonometric curve, reduced in proportion to the ordinates of the arbitrary curve which represents $\phi(x)$. This done, the area of the reduced curve taken from $x = 0$ to $x = \pi$ gives the exact value of the coefficient of $\sin x$ [i.e., b_1]; and whatever the given curve may be which corresponds to $\phi(x)$, whether we can assign to it an analytical equation, or whether it depends on no regular law, it is evident that it always serves to reduce in any manner whatever the trigonometric curve; so that the area of the reduced curve has, in all possible cases, a definite value, which is the value of the coefficient of $\sin x$ in the development of the function. [1822: 186.]

The existence of the definite integral of an arbitrary function was therefore based on the existence of the area of the set of ordinates of the function, a geometrically oriented viewpoint in keeping with the geometrical basis of the Euler-Fourier conception of an arbitrary function. Area as a quantity, however, was left undefined. The clarification and generalization of the notion of area is of fundamental importance to the historical development of the concept of the integral. Fourier's hypotheses (a) and (b) came to be regarded as lacking general validity (even granted the validity of (1)) until Lebesgue postulated a definition of area and of the integral in terms of which (a) and (b) follow as simple consequences of (1).

1.2 CAUCHY'S CONCEPTION OF THE INTEGRAL

Fourier thus remained faithful to the eighteenth-century usage of "discontinuous function," although his own work had actually undermined its basis by assigning an equation to every function. It was Augustin Cauchy (1789–1857) who is chiefly responsible for introducing the modern conception of continuity as well as for providing a precise definition of the definite integral as the limit of a sum. As early as 1814, and before he was familiar with Fourier's work, Cauchy came to recognize the importance of continuity in the modern sense and of the sum-conception of the definite integral. After 1818 when Cauchy learned of Fourier's work—particularly Fourier's interpretation of the coefficients a_n and b_n as integrals—he was undoubtedly all the more convinced of the importance of these notions.

In his *Cours d'analyse* [1821], Cauchy set forth the new conception of continuity: A function $f(x)$, uniquely defined and bounded for x between a and b $(a < b)$, is continuous between these limits if for such x "the numerical value of the difference $f(x+\alpha) - f(x)$ decreases indefinitely with that of α" [1821: 43]. It is interesting and probably significant that although continuity is a property of a function at a point, Cauchy makes fundamental the notion of continuity on an interval. On the other hand, discontinuity is presented as a property holding at a particular point: $f(x)$ is discontinuous at x_0 if it is not continuous in every interval about x_0. This difference in approach reflects Cauchy's conception of a discontinuous function, which we shall discuss further on.

Two years later, in his *Résumé des leçons données à l'École Royale Polytechnique sur le calcul infinitésimale* [1823], Cauchy defined the definite integral of a continuous function as follows: Let $f(x)$ be continuous for x in $[a,b]$, and consider a partition of this interval

$$a = x_0 < x_1 < x_2 < \cdots < x_{n-1} < x_n = b$$

and the Cauchy sum

$$S = \sum_{i=1}^{n} (x_i - x_{i-1}) f(x_{i-1}).$$

Using the continuity of $f(x)$ or, more precisely, the uniform continuity of $f(x)$—Cauchy confounded them—he showed that for any two partitions P and P', the corresponding sums S and S' could be made to differ by an arbitrarily small amount provided that the norms of P and P' are sufficiently small; and therefore "the value of S will obviously end up being constant . . . . This limit is what is called a definite integral" [1823: 125].

One of the principal advantages of this definition from Cauchy's point of view was that it enabled him to "demonstrate generally the existence of *integrals* or *primitive functions*" [1823: 9], whose various properties, he felt, could be studied only after a proof of their existence. For a function f continuous between a and b, Cauchy considered the function $F(x)$ defined by $F(x) = \int_a^x f.$ [1823: 151.] Since $\dfrac{F(x+h) - F(x)}{h} = \dfrac{1}{h}\int_x^{x+h} f = f(x+ch),$ for $|c| \leq 1$, F is not only continuous but differentiable as well, and Cauchy established the following results.

THEOREM I.   *F is a primitive function for f; that is, $F' = f$.*

THEOREM II.   *All primitive functions of f must be of the form $\int_a^x f + C$, where C denotes a constant; that is, if G is a function with a continuous derivative G', then $\int_a^x G' = G(x) - G(a)$.*

The following theorem was used by Cauchy to prove Theorem II.

THEOREM III.   *If G is a function such that $G'(x) = 0$ for all x in $[a,b]$, then $G(x)$ remains constant there.*

Theorems I, II, and III, as well as generalizations thereof, will henceforth be referred to collectively as the Fundamental Theorem of the Calculus or, more briefly, as the Fundamental Theorem. Its component theorems will be referred to as the Fundamental Theorem I, II, and III, respectively. The unsatisfactory revisions and limitations imposed upon the Fundamental Theorem after Riemann's extension of the integral concept to discontinuous functions are particularly significant for an appreciation of Lebesgue's work.

Cauchy's definitions of a continuous function and of the definite integral, being free from any reference to an analytical expression determining the function, apply without modification to the modern conception of a function as a correspondence $x \rightarrow f(x)$ between numbers x and $f(x)$. But Cauchy does not appear to have seriously entertained such a general notion. It is true that his definition of a function avoids reference to an analytical expression: he defined y to be a function of x if x and y are related in such a manner that a particular value of x determines the value of y. But, as Lebesgue has pointed out [1904: 4], Cauchy then proceeded to distinguish between explicit and implicit functions and described the latter as arising when the relations between x and y are given by equations that have not been algebraically resolved [Cauchy 1821: 32; 1823: 18]. Evidently, Cauchy still tended to regard functions as equations. (It was quite common to give a seemingly general definition so as to include both implicit and explicit functions.) That Cauchy did not seriously view functions in the modern sense is further suggested in a paper [1827b] that represents the first attempt to prove that the Fourier series of a function always converges. He attacked the problem by tacitly assuming that the function $f(x)$ retains a meaning when the complex variable $z = x + iy$ is substituted for x—an assumption, as Dirichlet was to point out, that has meaning only if $f(x)$ is regarded in terms of an analytical expression.

During the first half of the nineteenth century, a number of works, including one by Cauchy, were devoted to discussions of "discontinuous functions."[3] With the exception of Cauchy's, these works all reflect the influence of Fourier's notions and in particular his use of the term "discontinuous function." The fact that a function might be discontinuous in Cauchy's sense remained incidental and unnecessary. In modern notation the type of function these mathematicians had in mind can be represented in the form

$$(2) \qquad f(x) = \sum_{r=1}^{n} \chi_{I_r}(x) g_r(x),$$

where the domain $[a,b]$ of f is partitioned into subintervals $I_1, I_2, \cdots,$ I_n; the g_r are "continuous" functions (in the eighteenth-century sense of being determined by a single equation); and χ_{I_r} denotes the characteristic function of I_r (the characteristic function of a set S, χ_S, is defined: $\chi_S(x) = 1$ if x belongs to S; $\chi_S(x) = 0$ if x is not in S.) One author, combining the ideas of Fourier and Cauchy, adopted Cauchy's characterization of continuity and Fourier's conception of discontinuity![4]

Despite his adoption of the modern interpretation of continuity and discontinuity, Cauchy's conception of a discontinuous function had much in common with the functions described by (2). For example, in "Mémoire sur les fonctions discontinues" [1849], Cauchy used the term "discontinuous function" for those particular discontinuous functions arising in the solutions of the differential equations associated with physical problems. These functions, he assured his readers, are very well behaved: "It is important to observe that the discontinuous func-

3. Piola 1828; Libri 1831–33, 1842; Murphy 1833; Peacock 1833: 248ff; Rawson 1848; Boole 1848.
4. Piola 1828: 574, 579–81.

tions introduced into the calculus by consideration of the initial state only cease to be continuous for certain values [i.e., a finite number] of the variables involved" [1849: 123]. Cauchy proceeded to characterize these functions mathematically in exactly the same manner as in (2), the one difference being that the functions g_r were now continuous in his sense. Thus, although he may have recognized that the logical possibilities for discontinuous functions are not fully realized by this particular type, he was content to restrict his attention to them. Certainly his tendency to regard functions in terms of equations would have encouraged this attitude.

It is important to observe that Cauchy's theory of integration is entirely adequate for dealing with this special class of discontinuous functions; for, if f is bounded on $[a,b]$ and discontinuous at c in (a,b), then $\lim\limits_{\epsilon \to 0^+} \int_a^{c-\epsilon} f$ and $\lim\limits_{\epsilon \to 0^+} \int_{c+\epsilon}^b f$ necessarily exist and the definite integral can be defined as

$$(3) \qquad \int_a^b f = \lim_{\epsilon \to 0^+} \int_a^{c-\epsilon} f + \lim_{\epsilon \to 0^+} \int_{c+\epsilon}^b f.$$

(This is, in fact, the approach taken by Cauchy to extend his definition of the integral to unbounded functions.) The definite integral for a function with any finite number of discontinuities in $[a,b]$ can be defined analogously.

1.3   D I R I C H L E T ' S C O N T R I B U T I O N

It was not until the modern concept of a function was recognized and taken seriously that the problem of integrating discontinuous functions became apparent. For Fourier's and Cauchy's notion of a discontinuous function, Cauchy's definition of the integral entirely resolves the question of the meaning of the Fourier coefficients a_n and b_n as definite integrals. Discussion of this question was reopened, however, by Peter Gustav Lejeune-Dirichlet (1805–1859), who was the first mathematician to call attention to the existence of functions that are discontinuous on an infinite set of points in a finite interval and to the problem of extending the concept of the integral to functions of this nature.

Dirichlet's interest in the theory of Fourier series and integrals grew out of his personal acquaintance with Fourier during the years 1822–25, in which he continued his mathematical studies in Paris. He published [1829] the first rigorous proof that under certain general conditions the Fourier series of a function $f(x)$ converges. Dirichlet approached the problem by considering directly the behavior of the partial sums

$$S_n(x) = \tfrac{1}{2} a_0 + \sum_{k=1}^{n} a_k \cos kx + b_k \sin kx,$$

which he showed can be expressed in the closed form

$$S_n(x) = (1/\pi) \int_{-\pi}^{\pi} f(t) \, \frac{\sin \tfrac{1}{2}(2n+1)(t-x)}{\sin \tfrac{1}{2}(t-x)} \, dt.$$

Dirichlet explicitly assumed that the function f takes on a maximum or minimum value or is not continuous (in the sense of Cauchy) for at most a finite number of points. He then succeeded in demonstrating that $S_n(x)$ converges to $\tfrac{1}{2}[f(x-)+f(x+)]$ for x in $(-\pi,\pi)$ and to $\tfrac{1}{2}[f(-\pi+)+f(\pi-)]$ when x equals $-\pi$ or π. An examination of the proof, however, reveals that these conclusions about the limiting behavior of $S_n(x)$ are based upon the fact that f is monotonic in a sufficiently small interval about any x in $[-\pi,\pi]$ (with x possibly as an end point). Why, then, did Dirichlet include a continuity condition in his hypothesis? The answer is that he had to include a condition that would guarantee the existence of the definite integrals of f and of its products with $\cos nx$ and $\sin nx$ so that the Fourier coefficients a_n and b_n—and hence the integral representation of $S_n(x)$—would possess a meaning.

Dirichlet believed that the case of functions with an infinite number of extrema or points of discontinuity could be reduced to the case he had considered. The single necessary requirement upon the function, he claimed, would be that it possess a definite integral. As a sufficient condition for this for bounded functions he proposed the requirement that "if a and b denote two arbitrary quantities included between $-\pi$ and π, it is always possible to place other quantities r and s between a and b which are sufficiently close so that the function remains continuous in the interval from r to s" [1829: 169]. In modern terminology, Dirichlet's condition is that the set D of points of discontinuity of the function must be nowhere dense. At this time Dirichlet did not attempt to justify his claim, because a lucid proof "requires some details connected with the fundamental principles of infinitesimal analysis which will be presented in another note . . ." [1829: 169].

The promised note never came, and we are left to speculate on the manner in which he intended to extend the definition of the integral. It seems doubtful that he had in mind the approach later taken by Riemann, who extended Cauchy's definition by replacing the requirement that f be continuous with the weaker one that the Cauchy sums all converge to a unique limit. Had Dirichlet been thinking along these lines, he would have realized that his requirement of monotonicity suffices to guarantee the existence of the integral in this sense; he would not have included a continuity hypothesis in his theorem on convergence of Fourier series. A more likely interpretation is that Dirichlet had in mind extending the integral by generalizing equation (3) so as to cover the case in which D is an infinite set.

In his doctoral thesis [1864] Rudolf Lipschitz (1831–1904) attempted to extend Dirichlet's results on the convergence of Fourier series and, in the process, was led to consider a generalization of the integral based upon just such an interpretation. Although composed ten years after Riemann's researches, Lipschitz' paper was not influenced by them, and it appears doubtful that they were known to him. (Most mathematicians did not become aware of Riemann's paper until after its posthumous publication in 1867.)

Lipschitz began by listing the possible ways in which a function could fail to satisfy the conditions under which Dirichlet had proved his theorem. One way was if a bounded and piecewise monotonic function f was discontinuous at an infinite number of points in the interval $[-\pi,\pi]$. Lipschitz realized that Dirichlet's proof would apply if the integral concept could be extended to f, and he concluded that this could easily be accomplished for functions satisfying Dirichlet's condition on the set D of discontinuities (D is nowhere dense). The reason the extension seemed simple to him is that he believed that by "an appropriate argument" it could be shown that if D is nowhere dense, then D', the set of limit points of D, must be finite. If, for example, D' consists of the single point c in $(-\pi,\pi)$, then f has at most a finite number of points of discontinuity in $(-\pi,c-\epsilon)$ and in $(c+\epsilon,\pi)$, where ϵ is positive. Therefore $\displaystyle\int_{-\pi}^{c-\epsilon} f$ and $\displaystyle\int_{c+\epsilon}^{\pi} f$ can be defined as indicated in 1.2 (see equation 3), and $\displaystyle\int_{-\pi}^{\pi} f$ can then be defined as

$$\lim_{\epsilon\to 0^+} \int_{-\pi}^{c-\epsilon} f + \lim_{\epsilon\to 0^+} \int_{c+\epsilon}^{\pi} f.$$

These limits will always exist, since f is bounded.

Lipschitz' identification of nowhere dense sets D with sets such that D' is finite is not really surprising when one recalls that the theory of sets was nonexistent at this time. In particular, no interesting examples of nowhere dense sets had been constructed to keep pure reason from going astray. Thus, in connection with another infinite set E (the set of points for which $f(x)$ is a relative maximum or minimum), we find Lipschitz distinguishing three possibilities for the distribution of the points of E in $(-\pi,\pi)$: (1) There exists a point r in $[-\pi,\pi]$ with the property that outside of an interval about r there are at most a finite number of points of E; (2) Every subinterval of $(-\pi,\pi)$ contains an infinite number of points of E. (3) The interval $(-\pi,\pi)$ can be decomposed into a finite number of subintervals such that either (1) or (2) holds for each subinterval. An immediate consequence of this analysis is that E either possesses at most a finite number of limit points or is dense in some subinterval. For if the latter is not the case (i.e., E is nowhere dense), only intervals of type (1) can occur in the decomposition of $(-\pi,\pi)$ described in (3). In other words, this reasoning implies that D is nowhere dense if and only if D' is finite.

Despite these flaws, Lipschitz' paper represented an important contribution to the theory of trigonometric series. Because he slurred over the difficulties that do exist when D is nowhere dense, Lipschitz was primarily concerned with removing Dirichlet's monotonicity condition. He succeeded in proving Dirichlet's result with piecewise monotonicity replaced by what is now justifiably called a Lipschitz condition.

It is easily seen that the procedure of extending the integral proposed by Lipschitz can be applied to the case in which $D^{(2)} = (D')'$ is finite and, in general by induction, to the case in which $D^{(n)} = (D^{(n-1)})'$ is finite. This may have been what Dirichlet had in mind in 1829. In that case, he probably thought at that time that if D is nowhere dense, then $D^{(n)}$ must be finite for some n. (The derived sets $D^{(n)}$ were first introduced in print by Cantor [1872]; see 2.1.)

The tendency, suggested by Dirichlet's remarks, to regard "nowhere dense" as synonymous with "negligible for integration theory" and the tendency, evident in Lipschitz' paper, to conceive of a set as being dense either in the neighborhood of separated limit points or in an entire interval, continued until the early 1880's. Together these tendencies served to retard the introduction of a measure-theoretic viewpoint into the theory of integration for reasons that will be clearer further on.

Dirichlet probably found he was unable to carry out his proposed generalization of the integral: he never published his promised note, and when he published an elaborated version of his 1829 paper in German [1837], he assumed throughout the continuity of the functions under consideration and presented Cauchy's definition of the definite integral. However, we have with Dirichlet the beginnings of a distinction between the class of continuous functions and the class of integrable functions. It is also with Dirichlet that we find the first truly general conception of a function as any correspondence $x \rightarrow f(x)$ between real numbers. The idea of a function as a correspondence rather than as an analytical expression is, of course, suggested by Euler's and Fourier's conception of an arbitrary function as well as by Cauchy's definitions of continuity and the integral. But no one prior to Dirichlet appears to have taken the implications of this idea seriously. Indicative of this difference in attitude is Dirichlet's introduction of the now-familiar function which takes one value when x is rational and another when x is irrational: the function is defined neither by equations nor by a curve drawn—or drawable—in the plane. Dirichlet introduced it in his 1829 paper as an example of a function not satisfying his integrability condition and to which the integral concept could not be meaningfully extended. It was not until Lebesgue introduced his definition of the integral that Dirichlet was proved mistaken.

Prior to Riemann, Bernard Bolzano (1781–1848) appears to have been the only other mathematician to adopt the modern conception of a function and to consider the consequences of such an innovation. This is most clearly seen in his *Functionenlehre*, an unfinished work which remained largely unknown and in manuscript form until 1930.[5] The completed portions of this work are primarily concerned with the concepts of continuity and differentiation and have little to say about the integral.

1.4 RIEMANN

Bernhard Riemann (1826–66) probably acquired his interest in the problems connected with the theory of trigonometric series through contact with Dirichlet. After spending a year at Göttingen, Riemann went to Berlin, where he attended Dirichlet's lectures on number theory, the theory of definite integrals, and partial differential equations. Dirichlet quickly developed a personal interest in the young Riemann, who in turn considered Dirichlet the greatest living mathematician next to Gauss. Two years later Riemann returned to Göttingen and in 1851 presented his doctoral dissertation, "Grundlagen für eine allgemeine Theorie der Functionen einer veränderlichen complexen Grösse." The starting point of the "Grundlagen" is the substantially different definitions of the term "function" proposed by Riemann when the variable is real or complex. For a correspondence $w = f(z)$ between complex variables, it was required that $f'(z)$ exist. But for real variables, Riemann's definition reflects the influence of Dirichlet, a function being defined as any correspondence. And because Riemann believed, as did

5. B. Bolzano, *Functionenlehre*, Vol. 1, *Schriften* (Prague, 1930).

Dirichlet, that every continuous function of a real variable can be represented by its Fourier series,[6] he concluded that the definition of a real-valued function as an analytical expression is coextensive with the ostensibly more general one in terms of arbitrary correspondences.

For his *Habilitationsschrift* of 1854 Riemann decided to undertake the study of the representation of functions by trigonometric series [Riemann 1902: 227–65]. The first part consists of a history of the subject, in the preparation of which he received first-hand information from Dirichlet. After discussing Dirichlet's own contribution, Riemann remarked that it is reasonable to assume "that the functions not covered by Dirichlet's analysis do not occur in nature" [1902:237]. Nevertheless, for two reasons he felt it worthwhile to consider the case of more general functions. First because, as Dirichlet himself had maintained, "this subject is very closely related to the principles of the infinitesimal calculus and can help bring greater clarity and precision to these principles" [1902: 238]. And, secondly, because "the application of Fourier series is not limited to physical investigations; it is now being applied successfully in a domain of pure mathematics, the theory of numbers, and here those functions whose representability by a trigonometric series Dirichlet did not investigate seem to be of importance" [1902: 238].

Riemann began his own investigation with the question left unresolved by Dirichlet: In what cases is a function integrable? But Riemann interpreted the meaning of integrability in terms of the Cauchy sums corresponding to the function: Under what conditions on the function $f(x)$ do these sums approach a unique limit as the norm of the corresponding partitions approaches zero? He took it for granted that this will be the case if and only if

$$(R_1) \qquad \lim_{||P|| \to 0} (D_1\delta_1 + D_2\delta_2 + \cdots + D_n\delta_n) = 0,$$

where the δ_i denote the lengths of the subintervals of $[a,b]$ formed by the partition P, and the D_i denote the oscillation of $f(x)$ on the corresponding subintervals. The common limit of the Cauchy sums was by definition $\int_a^b f(x)dx$.

Next Riemann showed that the condition for integrability (R_1) can be replaced by an equivalent condition. For a fixed number $d > 0$ let

6. Gauss also believed this. (See Gauss, *Werke*, Vol. 7, 2nd ed. (Leipzig, 1906), pp. 470–72.) Paul du Bois-Reymond was the first to show that this is not true in general. (See [du Bois-Reymond 1873; 1876: 75–76, 79–100].)

$\Delta = \Delta(d)$ denote the maximum of all sums $D_1\delta_1 + D_2\delta_2 + \cdots + D_n\delta_n$ corresponding to partitions of norm less than or equal to d. Then f is integrable if and only if Δ converges to 0 with d. For $\sigma > 0$ a second given quantity and for any partition P of norm less than or equal to d, let $s = s(P,\sigma)$ denote the sum of the δ_i for which D_i is greater than σ. The contribution of these intervals to the sum $D_1\delta_1 + D_2\delta_2 + \cdots + D_n\delta_n$ is at least σs. Therefore, $\sigma s \leq \delta_1 D_1 + \delta_2 D_2 + \cdots + D_n\delta_n \leq \Delta$, and $s \leq \Delta/\sigma$. Consequently, if f is integrable, s converges to 0 with d; that is:

> (R₂)   *Corresponding to every pair of positive numbers ϵ and σ there exists a positive d such that if P is any partition of norm less than or equal to d, then $s(P,\sigma)$ is less than ϵ.*

Riemann showed that (R₂) is also a sufficient condition for the integrability of f: if f satisfies (R₂) and $\epsilon > 0$, $\sigma > 0$ are given, there exists a $d > 0$ satisfying (R₂); thus, for all partitions P such that $\|P\| < d$,

$$D_1\delta_1 + D_2\delta_2 + \cdots + D_n\delta_n \leq Ds + \sigma(b - a) \leq D\epsilon + \sigma(b - a),$$

where D denotes the oscillation of f on $[a,b]$. (R₁) then follows, since ϵ and σ can be taken arbitrarily small.

The reader will recognize in conditions (R₁) and (R₂) the germs of the concepts of (Jordan) measurability and outer content, respectively, each of which played a fundamental role in bringing about the creation of the modern theory of integration. It must be kept in mind, however, that Riemann himself did not regard (R₁) and (R₂) in terms of these concepts. It was not for some time to come that measure-theoretic ideas were explicitly introduced.

The concept of an integrable function thus comes to the fore with Riemann and is established independently of any continuity considerations, although it was soon discovered that Riemann integrability entails certain continuity properties. Riemann also pointed out that the class of integrable functions includes functions possessing an infinity of points of discontinuity in every interval. And "since these functions are as yet nowhere considered, it will be good to begin with a specific example" [1902: 242]. The example is defined in the following manner. For every real number x let (x) denote (a) $x - m(x)$ if $x \neq n/2$, for n odd, where $m(x)$ is the integer such that $|x - m(x)|$ is a minimum, or (b) 0 if $x = n/2$, for n odd. The function $y = (x)$ is thus discontinuous at the points $x = n/2$ with n odd. (See Figure 1.) Riemann then defined the function

$$f(x) = (x) + \frac{(2x)}{2^2} + \cdots + \frac{(nx)}{n^2} + \cdots$$

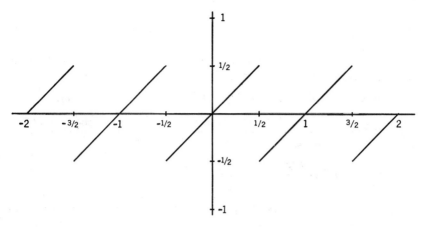

Fig. 1 $y = (x)$ for $-2 \leq x \leq 2$

As one would suspect, this function is discontinuous at any point of the form $x = m/2n$, where m and $2n$ are relatively prime. In fact, for these values of x, the right-hand and left-hand limits of f at x are:

$$f(x+) = f(x) - (1/2n^2)\left[\sum_{k=0}^{\infty} 1/(2k+1)^2 \right] = f(x) - (\pi^2/16n^2);$$

$$f(x-) = f(x) + (1/2n^2)\left[\sum_{k=0}^{\infty} 1/(2k+1)^2 \right] = f(x) + (\pi^2/16n^2).$$

Thus f is discontinuous on a dense subset of the real numbers. Nevertheless, f is integrable on any interval because for any $\sigma > 0$ there are only a finite number of points $x = m/2n$ at which the jump, $f(x-) - f(x+) = \pi^2/8n^2$, is greater than σ. Condition (R_2) is consequently satisfied.

    The originality of Riemann's genius is reflected in his radically different approach to the representation of functions by trigonometric series. He begins by considering a trigonometric series

$$\Omega(x) = A_0 + A_1(x) + A_2(x) + \cdots + A_n(x) + \cdots,$$

where $A_0 = \frac{1}{2}a_0$; $A_n(x) = a_n \cos nx + b_n \sin nx$; and $A_n(x)$ converges to 0 (uniformly in x)[7] as n increases. The series obtained by formally integrating Ω term-by-term twice,

    7.   Riemann did not explicitly distinguish between uniform and non-uniform convergence, but he assumed throughout that the $A_n(x)$ converge uniformly.

$$(4) \qquad F(x) = C + C'x + A_0 x^2/2 - \sum_{n=1}^{\infty} A_n(x)/n^2,$$

converges for every value of x and represents a continuous function of x. Furthermore, he showed that F has two important properties:

$\mathfrak{D}^2 F(x)$ exists and equals $\Omega(x)$ whenever Ω converges, where

$$(5) \quad \mathfrak{D}^2 F(x) = \lim_{a,b \to 0} \frac{F(x+a+b) - F(x+a-b) - F(x-a+b) + F(x-a-b)}{4ab}.$$

$$(6) \quad \text{For all } x, \lim_{a \to 0} \frac{F(x+a) - 2F(x) + F(x-a)}{a} = 0.$$

Using these results Riemann was able to obtain necessary and sufficient conditions (in terms of F) for the representability of a function at a point by a trigonometric series Ω. What is particularly important about this part of his study is that no assumption is made about the nature of the coefficients a_n and b_n occurring in Ω; Ω need not be a Fourier series, that is, a trigonometric series in which the coefficients are given by Fourier's integral formulae. The distinction between trigonometric and Fourier series was first introduced by Riemann, who also provided a number of examples of trigonometric series which are not Fourier series. This distinction becomes important in connection with the investigations of Cantor on the theory of sets.

Riemann's examples raised more problems than he was able to resolve, and it might have been the incompleteness of his results that caused him to leave them unpublished during his lifetime. It was not until 1867 that Richard Dedekind had the work included in the *Abhandlungen der Königlichen Gesellschaft der Wissenschaften zu Göttingen*, considering its publication justified not only because of the interesting nature of the subject but also because of its treatment of "the most important principles of infinitesimal analysis . . ." [Riemann 1902: 227n]. Dedekind's judgment was soon borne out by the research activity stimulated by Riemann's ideas.

2

The Development
of Riemann's Ideas: 1870–80

2.1 TRIGONOMETRIC SERIES AND TERM-BY-TERM INTEGRATION

Fourier had taken for granted the permissibility of term-by-term integration of infinite series of functions. During the first half of the nineteenth century a more rigorous analysis of series was developed, notably in the works of Cauchy, Gauss, and Abel. But the emphasis was put primarily upon questions of convergence and not upon the interchangeability of infinite summation with other limit processes (as, for example, when a series is integrated by summing the integrals of its terms). Thus, the validity of term-by-term integration of continuous functions was posited by Cauchy [1823 : 237] and also assumed by Gauss in his attempt to prove the (false) proposition that every continuous function is representable by its Fourier series.[1]

The validity of interchanging limit processes was called into question by N. H. Abel (1802–29) in 1826. Cauchy [1821 : 120] had included in his *Cours d'analyse* the proposition that if a series of functions converges in the neighborhood of a point a and if its terms are continuous at a, then the series converges to a function that is likewise continuous at a. Abel [1826] used the example $\sum_{n=1}^{\infty} (-1)^{n+1}(\sin nx)/n$ to illustrate the incorrectness of Cauchy's proposition. He also showed, however, that for certain types of infinite series, continuity of the sum can be

---

1. K. Gauss, *Werke*, Vol. 7, 2nd ed. (Leipzig, 1906), pp. 470–72.

deduced from the continuity of the terms; it is precisely the uniform convergence of these series to which he appeals. But Abel did not extricate this property from the particular class of series under consideration and recognize its general significance.

That step was taken by Karl Weierstrass (1815–97), who may have been influenced by Abel's suggestive remarks. At least as early as 1841 Weierstrass was aware of the theoretical importance of the distinction between uniform and nonuniform convergence.[2] After he accepted the offer of a university position at Berlin in 1856, he stressed the importance of uniform convergence in his lectures; in particular he showed that the validity of term-by-term integration can be demonstrated for any series which converges uniformly on the interval over which it is integrated.

It should be noted in passing that Philipp Seidel (1821–1896) and George Stokes (1819–1903) had discovered independently that if a series of continuous functions does not converge "infinitely slowly" at a point a—i.e., the series converges uniformly in a neighborhood of a—then the series will be continuous at a [Seidel 1850; Stokes 1848].[3] But they failed to relate their ideas to the term-by-term integration of infinite series, probably because they were concerned, in effect, with the property of uniform convergence in the neighborhood of a particular point and not as a global property. Stokes, in fact, explicitly accepted Cauchy's proposition on term-by-term integration [1880: I, 242, 255, 268, 283]. And when Cauchy [1853] rectified his original statement by adding the requirement of uniform convergence, he still failed to recognize the equally untenable nature of his proposition on term-by-term integration. It was Weierstrass who was primarily responsible for raising serious doubts among mathematicians about the general validity of term-by-term integration by restricting its demonstration to uniformly converging series.

Outside the circle of Weierstrass' students, the significance of uniform convergence for the term-by-term integration of series was not generally recognized until Heinrich Heine (1821–81) called attention to it dramatically in a paper [1870] on trigonometric series. Heine, who was a professor of mathematics at the University of Halle, probably became familiar with Weierstrass' ideas through Georg Cantor (1845–1918), who had come to Halle after receiving his doctorate at Berlin in 1867.

Fourier had based the existence and uniqueness of a trigonometric

2. Weierstrass made this distinction in a paper written in 1841 but first published in 1894 [Weierstrass 1894–1927: I, 67–74].

3. Stokes's conception of noninfinitely slow convergence is actually not entirely free from ambiguities. See [G. H. Hardy 1918].

series representation of an arbitrary function upon the permissibility of term-by-term integration. Now that this was in doubt for nonuniformly convergent series, the question arises, Heine observed, as to whether the coefficients in the representation of a function f in the form

$$(1) \qquad f(x) = \tfrac{1}{2} a_0 + \sum_{n=1}^{\infty} a_n \cos nx + b_n \sin nx$$

are necessarily unique. The problem posed by Heine is closely related to, and was probably suggested in part by, Riemann's disclosure of the existence of trigonometric series which are not Fourier series.

As Heine observed, the Fourier series of a continuous function that satisfies Dirichlet's condition of piecewise monotonicity must always converge uniformly but the convergence is necessarily nonuniform in the neighborhood of a point such that $f(x+)\neq f(x-)$. In this case, however, the Fourier series of f (still assumed to satisfy Dirichlet's conditions) converges to $f(x)$ uniformly "in general." A series $\sum_{n=1}^{\infty} f_n(x)$, defined for x in $[a,b]$, is said to converge uniformly "in general" to $f(x)$ in $[a,b]$ if there exists a finite number of points $x_1, x_2, \cdots, x_n$ in $[a,b]$ such that the series converges uniformly to $f(x)$ on any subinterval of $[a,b]$ not containing the x_i. Heine proposed to prove that a function f can be represented in at most one way by a trigonometric series which converges uniformly in general. By definition, this type of convergence admits the possibility that the series does not represent $f(x)$ for $x=x_1$, $x_2, \cdots, x_n$, and Heine tells us it was Cantor who induced him to consider this possibility.

By considering the difference of two possibly distinct representations of f of the form (1) Heine reduced the problem to the following: If, for all x except those forming a set P,

$$(2) \qquad \tfrac{1}{2} a_0 + \sum_{n=1}^{\infty} a_n \cos nx + b_n \sin nx = 0,$$

then are the coefficients a_n and b_n all equal to zero? He was able to show that the answer is Yes when the convergence in (2) is uniform in general with respect to the set P, which is thus taken to be finite. Using this hypothesis, he first showed that a_n and b_n converge to zero as n increases; this result then insures the existence of Riemann's function $F(x) = \tfrac{1}{2} A_0 x^2 + \sum_{n=1}^{\infty} -A_n(x)/n^2$. Heine's idea was to use F and its properties, as demonstrated by Riemann, to show that a_n and b_n are actually zero.

Cantor was induced by Heine's research to study the uniqueness problem, and he set himself the task of removing the hypothesis of uniform convergence in general. This he accomplished in a series of papers of 1870–71 [Cantor 1870a; 1871a]. Having thus completely removed the assumption of uniform convergence in general and thereby the reason for restricting P to be finite, Cantor [1872] next turned to the question of uniqueness when P is an infinite set.

At this time no one had carefully analyzed the properties of infinite sets, although dense and nowhere dense sets had entered into the work of Dirichlet, Riemann, Lipschitz, and, to a larger degree, Hermann Hankel (2.2), as well as into discussions of the differentiability of continuous functions (2.3). Cantor consequently began with some basic definitions. A point x is a limit point of a set of points P if every neighborhood of x contains infinitely many points of P. (By a neighborhood of x, Cantor meant an interval with x in its interior.) The first derived set of P, denoted by P', consists of all limit points of P. If P' is not finite, then it has a derived set (P is assumed bounded) which is denoted by P'' or $P^{(2)}$ and called the second derived set of P. In this manner, the nth derived set of P, written as $P^{(n)}$, can be defined recursively as $(P^{(n-1)})'$. Finally, a set P is defined to be of the nth type if $P^{(n)}$ is a finite set. To illustrate these notions, Cantor considered the set of all rational numbers between 0 and 1; since its derived set contains all real numbers between 0 and 1, it is not of any type n. He also considered the set of points of the form $1/n$ where n is a positive integer; in this case the derived set contains the single number 0 and the set is of type 1. These simple examples—the only ones given by Cantor—reflect the limited number of actual examples of infinite sets known to mathematicians at this time.

Cantor's principal result is that if (2) holds except on a set P of finite type n (later termed a set of the first species by Cantor), then the coefficients in (2) are all zero. The proof derives from Heine's proof as modified by Cantor to establish the theorem for $n=0$; Cantor realized that the proof can be extended by induction to any integer n. The idea is similar to that employed by Lipschitz (1.3), and it is possible that his paper influenced Cantor. But Cantor did not make Lipschitz' mistake of thinking that sets of types 0 and 1 exhaust the possibilities of nowhere dense sets. He realized that there must exist sets of any type n (which are easily seen to be nowhere dense).[4]

In the development of the theory of integration, these papers by Cantor and Heine are doubly significant: their publications called at-

---

4.   Cantor's belief in the existence of these sets and the conception of a set of type n itself are closely connected with his theory of real numbers. See [Jourdain 1910: 23–26, esp. 24n1, 26n2].

tention to the fact that term-by-term integration is a problem, and also marked the beginning of the development of the theory of infinite sets, a theory which greatly influenced integration theory.

The work of Cantor and Heine had led to the discovery that the co-efficients in a trigonometric series representation are unique but had not touched upon the nature of the coefficients. In April of 1872 two Italian mathematicians, Giulio Ascoli and Ulisse Dini, submitted papers dealing with this point. Ascoli [1873] considered a function which is continuous except at a finite number of points, and Dini [1874] a function whose points of discontinuity possess at most a finite number of limit points. They succeeded in showing that for these functions the co-efficients in a trigonometric series representation must be the Fourier coefficients.

The method of proof adopted by both Ascoli and Dini is based upon showing that Riemann's function (1.4, formula (4))

$$F(x) = \tfrac{1}{4} a_0 x^2 - \sum_{n=1}^{\infty} (a_n \cos nx + b_n \sin nx)/n^2$$

can be represented in the form

$$F(x) = \int_{-\pi}^{x} dt \int_{-\pi}^{t} f(y) dy + cx + d.$$

Once this has been established, the nature of the coefficients can then be deduced by integrating $F(x)$ term-by-term and applying integration by parts to $G(x) = \int_{-\pi}^{x} dt \int_{-\pi}^{t} f(y) dy.$

In a paper of 1874, Paul du Bois–Reymond (1831–89) considered the more general case of a function f which is simply assumed to be integrable in Riemann's sense [1875b]. For $\Phi(x) = F(x) - G(x)$, the Fundamental Theorem I together with a theorem due to Riemann (formula (5) of 1.4) implies that $\mathfrak{D}^2\Phi(x) = 0$, where

$$\mathfrak{D}^2\Phi(x) = \lim_{\alpha \to 0} \Delta^2\Phi(x)/\alpha^2 = \lim_{\alpha \to 0} [\Phi(x + \alpha) + \Phi(x - \alpha) - 2\Phi(x)]/\alpha^2.$$

If the points of discontinuity form a first species set (that is, a set of type n), then it follows that $\Phi(x) = cx + d$ (this results from a lemma of H. A. Schwarz's that Cantor had called to attention in his researches on the uniqueness of trigonometric series representations). But du Bois–Reymond was faced with the fact that a Riemann-integrable function may be discontinuous on a dense set of points.

In order to circumvent this difficulty, du Bois–Reymond introduced a new concept: the aggregate of values (*Wertevorrat*) of a function at a point. The origins of this concept can be seen in a paper on infinite series $\sum_{k=1}^{\infty} u_k$ with bounded, but not necessarily convergent, partial sums $S_n = \sum_{k=1}^{n} u_k$. Here du Bois-Reymond [1870: 3ff] introduced what he called the "limits of indetermination" (*Unbestimmtheitsgrenzen*) of the series, now denoted by $\liminf_{n \to \infty} S_n$ and $\limsup_{n \to \infty} S_n$.[5] These he regarded as the lower and upper bounds of all limiting values of the sequence S_n. Conceived in this manner, these notions readily generalize to functions. Thus, the aggregate of values of f at x_1 is defined to be the set of all limiting values of $f(x)$ as x approaches x_1. The greatest lower bound and the least upper bound of this set are then defined to be the limits of indetermination of f at x_1. The difference of these limits is similar to what is now termed the oscillation of f at x_1, $\omega_f(x_1)$.

Thus, du Bois-Reymond was able to write

$$(3) \qquad \lim_{\alpha \to 0} \Delta^2 \Phi(x)/\alpha^2 = f(x) - r,$$

where r is between the limits of indetermination of f at x. The sense of (3) is, in more modern notation, that

$$f(x) - \limsup_{y \to x} f(y) \leq \liminf_{\alpha \to 0} \frac{\Delta^2 \Phi(x)}{\alpha^2}$$

$$\leq \limsup_{\alpha \to 0} \frac{\Delta^2 \Phi(x)}{\alpha^2}$$

$$\leq f(x) - \liminf_{y \to x} f(y).$$

From equation (3), by applying Riemann's integrability condition (R_1) and the lemma due to Schwarz, he was then able to show that Φ is a linear function. The resulting proof is extremely long and cumbersome and stands in sharp contrast to the proof later given by Lebesgue (6.1). Lebesgue's proof applies to any bounded function and, ironically, is based upon the fact that term-by-term integration is a valid operation upon series that are uniformly bounded, whether or not they converge uniformly. At the time, however, du Bois–Reymond's result represented a triumph for Riemann's theory of integration, and it appeared that Fourier's original proposition—that the only possible trigonometric series representation of an arbitrary bounded function is

---

5. Gauss had already considered these limits in an unpublished work written about 1800–01. See [Schlesinger 1912: 136–40].

its Fourier series—could not be maintained without further restricting the nature of the function.

The mathematician chiefly responsible for making Riemann's ideas known in France was Gaston Darboux (1842–1917). In 1870 he helped found the *Bulletin des sciences mathématiques et astronomiques*, and the first volume contained a lengthy, favorable review of a work by Hermann Hankel that developed Riemann's ideas. Then in 1873 Riemann's paper was translated by Darboux and published in the *Bulletin*. Actually Darboux did more than translate this "beau Mémoire." He studied it assiduously and in 1873, before the Société Mathématique de France, read two papers which dealt with the integrals of discontinuous functions, continuous functions without derivatives, and the theory of functions.[6] Darboux's researches culminated in a "Mémoire sur la théorie des fonctions discontinues" [1875]. One cannot read this work without being impressed by its lucidity and rigor; among the leading French mathematicians (with the exception of Ossian Bonnet), this degree of precision was uncommon. But Riemann's introduction of a radically different class of functions had made the need for rigor acute. "Many points, which could rightfully be regarded as evident or which could be conceded in the applications of science to the usual functions, must be submitted to a rigorous examination in the statement of propositions relative to more general functions" [Darboux 1875: 58].

A large portion of Darboux's memoir is devoted to establishing the basic propositions of the theory of functions of real variable and to constructing functions with exceptional properties. But Darboux [1884: 29] characterized his memoir as "in reality a study of the principles on which the integral calculus rests." Of particular relevance is Darboux's discussion of the integration of infinite series. He was familiar with the papers of Heine and Cantor and supplied an elegant proof which established both the integrability of a uniformly convergent series of integrable functions and the validity of term-by-term integration.

To indicate what can happen when a series converges nonuniformly, Darboux introduced the series $\sum_{n=1}^{\infty} u_n(x)$, where

$$u_n(x) = -2n^2xe^{-n^2x^2} + 2(n+1)^2xe^{-(n+1)^2x^2}.$$

This series converges to the continuous function $-2xe^{-x^2}$, which, when integrated over $[0,x]$, yields

$$\int_0^x (-2te^{-t^2})dt = e^{-x^2} - 1.$$

6. The contents are briefly described in *Paris, Soc. Math. Bull., 1* (1872–73), 121; *2* (1873–74), 66.

But term-by-term integration of the series over $[0,x]$ yields

$$\sum_{n=1}^{\infty} \int_0^x u_n(t)dt = \sum_{n=1}^{\infty} \left[e^{-n^2 x^2} - e^{-(n+1)^2 x^2} \right] = e^{-x^2}.$$

The series fails to converge uniformly because, as Darboux realized, $\left| R_n(1/n) \right| = 2n/e$, where $R_n(x) = \sum_{k=n}^{\infty} u_k(x)$. In other words, the series is not uniformly bounded on $[0,x]$. (A series $u(x) = \sum_{n=1}^{\infty} u_k(x)$ or, equivalently, the sequence $S_n(x) = \sum_{k=1}^{n} u_k(x)$, is said to be uniformly bounded on $[a,b]$ if there exists a number B such that $\left| S_n(x) \right| \le B$ for all x in $[a,b]$ and all n.) But Darboux did not introduce the notion of uniform boundedness and apparently had no idea of its importance for term-by-term integration. Heine, du Bois-Reymond, and Darboux all appear to have been so impressed by the discovery of the importance of uniform convergence that they paid little attention to the possibility of being able to make positive statements about nonuniformly converging series.

2.2   THE REFORMULATION OF RIEMANN'S INTEGRABILITY CRITERIA

Riemann had expressed his integrability condition in two forms, (R_1) and (R_2). It was the second form that eventually led to the introduction of the notion of outer content and the beginning of a measure-theoretic viewpoint within integration theory. During the 1870's, the measure-theoretic implications of (R_2) were not developed, even though the work of Hermann Hankel made these implications much more visible by showing that the integrability of a function depends on the nature of certain sets of points. An explanation for this phenomenon is the confusion that prevailed during this period over the natures of three types of "negligible" sets: Cantor's first species sets (sets of finite type n); nowhere dense sets; and sets that can be enclosed in a finite number of intervals of arbitrarily small total length, that is, sets of zero content. These types of sets are all distinct, but this fact was not generally recognized in the 1870's. As a result, the special importance of a measure-theoretic characterization of negligibility as opposed to a "topological" one was not appreciated. The discovery in the 1880's of the distinctness of these types of negligible sets was to provide the fillip for the development of the theory of outer content (see Chapter 3). In a sense, the limited conception of a nowhere dense set that was held by Dirichlet in 1829 (apparently), by Lipschitz, and by Hankel is embodied in Cantor's notion of a first species set. For this very reason Cantor's introduction of first species sets did not help clarify the distinct natures of nowhere dense sets and sets of zero content; a first species set is unfortunately both nowhere dense and of zero content.

Riemann's memoir passed on to his successors the viewpoint that a function of a real variable is to be regarded as any correspondence $x \rightarrow f(x)$ between real numbers; a function need not possess the properties (e.g., continuity and integrability) usually ascribed to the more familiar examples of functions occurring in analysis, even though, as in his examples, the function may be represented by an analytical expression. One of his disciples, Hermann Hankel (1839–73), a professor of mathematics at the University of Tübingen, was particularly impressed by Riemann's viewpoint and discoursed upon it at length in two works composed at about the same time—an article "Grenze" in the *Allgemeine Encyklopädie der Wissenschaften und Künste* [1871: 189–206] and a lengthy essay, "Untersuchungen über die unendlich oft oszillierenden und unstetigen Funktionen" [1870].

Hankel made explicit what was implicit with Dirichlet and Riemann: functions as conceived by these mathematicians do not possess general properties. Therefore, the first step in an analysis of the new function concept should be to remove all ideas about functions connected with the older notions and to analyze "the multiplicity of possible relationships between two variables" that are allowed by Dirichlet's concept [1870: 68]. The outcome of this analysis was Hankel's classification of functions. The first class consists of all continuous functions; the second, closely resembling the first in terms of the properties of its members, of all functions continuous except at a finite number of points in any finite interval. For Euler, Fourier, and Cauchy these two classes would have exhausted the possibilities of the function concept. But Dirichlet and Riemann had called attention to the possibility of functions with an infinite number of discontinuities in a finite interval. These functions are further divided by Hankel into two classes, which he refers to as the class of pointwise discontinuous functions (*punktiert unstetige Funktionen*) and the class of totally discontinuous functions.

The basis for this subdivision was suggested by the distinction emphasized by Dirichlet and Riemann between integrable and nonintegrable discontinuous functions. But Hankel added some important ideas of his own, such as the concept of the jump of a function at a point [1870: 87ff]. Riemann had used this term in connection with his proof that $f(x) = \sum_{n=1}^{\infty} (nx)/n^2$ is integrable (1.4), but Hankel modified the term to apply to any function, not simply to those possessing one-sided limits $f(x-)$ and $f(x+)$ at x. Thus, Hankel defined a function to make jumps greater than the positive number σ at x if for every positive ϵ there exists a δ such that $|\delta| < \epsilon$ and $|f(x+\delta) - f(x)| > \sigma$. The jump of f at x is then defined as the "largest" σ for which f makes jumps greater than σ at x. (More precisely, the jump of f at x is the least upper bound of such σ.) This notion is, of course, the forerunner of the more familiar notion of the oscillation of a function at a point.

The distinction between pointwise discontinuous and totally discontinuous functions is then formulated in the following manner. Let S_σ denote the set of points x at which f makes jumps greater than σ. If, for a discontinuous function, S_σ is nowhere dense for every positive σ—in Hankel's terminology, the points of S_σ lie "scattered" (*zerstreut*) on the interval of definition of the function—then the function is called pointwise discontinuous. Hankel regarded these functions as forming the bridge between the first two classes of functions and the remaining discontinuous functions. The latter, the totally discontinuous functions, have the property that, for some σ, the points of S_σ "fill up" (*erfüllen*) an entire interval; that is, S_σ is dense in some interval.

By making the distinction between pointwise and totally discontinuous functions, Hankel believed he had separated the functions amenable to mathematical analysis from those beyond its reaches: A function, he thought, would be Riemann-integrable if and only if it were pointwise discontinuous. More specifically, Hankel thought he had established the following result: A function is pointwise discontinuous if and only if for every positive σ the total magnitude of the intervals in which the oscillation of the function is greater than σ can be made arbitrarily small. It is not difficult to see that the "if" part of the statement is in fact correct: if S_σ is dense in an interval I for some value of σ, the oscillation of the function in any interval overlapping I will be greater than σ. Hence, the total magnitude of the intervals (of a partition of the fundamental interval $[a, b]$) in which the oscillation of the function is greater than σ will always be greater or equal to the length of I, and the function will not be integrable. It is the "only if" part—which in effect asserts that a nowhere dense set must have zero content—that is erroneous.

Before discussing Hankel's proof of this part of the proposition, it will be helpful to consider the actual examples of pointwise and totally discontinuous functions that were known to him, for they reveal the degree to which he had apprehended the possibilities for the various types of sets occurring in his definitions.

Hankel generalized Dirichlet's example in the following manner. Let

$f(x) = 1$ for all x in $[0,1]$ except for x in the infinitely many intervals with $1/2^n$ as center and of length ζ^n for a fixed ζ, $0 < \zeta < 1$, and for $n = 1, 2, 3, \cdots$. In these intervals f possesses the erratic behavior of Dirichlet's function, being equal to 1 for irrational x and equal to 0 for x rational. "The total magnitude s of the intervals throughout which oscillations equal to 1 occur is $s = \zeta + \zeta^2 + \zeta^3 + \cdots = \zeta/(1 - \zeta)$" [1870: 86]. In other words, f is totally discontinuous. The following example is particularly significant for understanding the basis of Hankel's remarks about pointwise discontinuous functions. Define f as follows: $f(1/2^n) = 0$ for $n = 1, 2, 3, \cdots$, and $f(x) = 1$ for all other values of x. At the infinite number of points $1/2$, $1/2^2$, $1/2^3$, $\cdots$, f makes jumps of magnitude 1. But, in contrast to the other example, there is no interval which is "filled up" with points of discontinuity. Indeed: "If an interval containing the point $x = 1/2^n$ is required to enclose only the discontinuity occurring at this point, it can be taken arbitrarily small. Moreover, the total magnitude s of all the intervals of this kind can be made arbitrarily small. For if the magnitude of the interval enclosing $x = 1/2^n$ is taken to be ϵ^n, then their sum $s = \epsilon + \epsilon^2 + \epsilon^3 + \cdots = \epsilon/(1 - \epsilon)$ becomes arbitrarily small with ϵ" [1870: 86].

With this example in mind, consider now Hankel's proof that if S_σ is nowhere dense, it can be enclosed in intervals of arbitrarily small total length. If S_σ contains a finite number of points, Riemann's proof for $f(x) = \sum_{n=1}^{\infty} (nx)/n^2$ can be repeated.

> In the case in which the number of these points is infinitely large . . . one can proceed as follows: Divide the entire interval under consideration into intervals each of which surrounds one of these discontinuity points with jumps greater than σ, and choose these intervals so large that together they fill up the entire interval. If each of these intervals is then imagined reduced to the nth part but in such a way that it still surrounds the corresponding point of discontinuity, then the remaining part of the interval is free from jumps greater than σ; the sum of those intervals in which jumps greater than σ occur, however, is the nth part of the whole interval and therefore can be made arbitrarily small by increasing n. q.e.d. [1870: 87–88.]

We can dispel the ambiguity surrounding this proof by considering another erroneous demonstration of his to the effect that the limits $f(x+)$ and $f(x-)$ must always exist when $f(x)$ is pointwise discontinuous. It begins: "Fix a point $x = a$ at which there are jumps greater than σ; then look for *the next point* $x = a + h$ at which jumps greater than σ occur; in the interval lying between them the oscillations are less than 2σ . . . " [Hankel 1870: 90; italics have been added]. Here Hankel's actual understanding—as opposed to his formal definition—of a "scattered" set becomes more evident: as conceived by Hankel, a "scattered" set does not contain any limit points. And, as the rest of this proof indicates, the points of S_σ cannot be limit points of any of the sets $S_{\sigma'}$ for σ' less than σ. This strongly suggests that Hankel had the set $\{1/2^n\}$ in mind and thought all nowhere dense sets would be similar enough to it to share those properties he had assumed in his arguments.

In summary, Hankel confounded "topologically negligible" sets (i.e., nowhere dense sets) with sets that are negligible from a measure-theoretic point of view. In attempting to topologically characterize the sets S_σ corresponding to an integrable function, he was following in the direction suggested by Dirichlet. And, as was the case with Dirichlet and Lipschitz, it was the inadequacy of his understanding of the possibilities of infinite sets—in particular, nowhere dense sets—that led him astray. It was not until it was discovered that nowhere dense sets can have positive outer content that the importance of negligible sets in the measure-theoretic sense was recognized. Thus the idea of a theory of measure implicit in Riemann's integrability condition (R_2) remained obfuscated by the prominence given to topological ideas that were seemingly equivalent. Nevertheless, by introducing the notion of the jump of a function at a point, Hankel had focused attention upon the properties of sets (the S_σ); it still remained to be recognized that the measure-theoretic properties of sets are crucial for the theory of integration.

Riemann had extended Cauchy's definition of the integral by recognizing the nonessential nature of the requirement that the integrand be continuous. Hankel discovered, however, that the points of continuity of an integrable function must form a dense set [1870: 90]. The proof is based upon the fact that the sets S_σ must be nowhere dense for every positive value of σ. Thus, for I_0 any specified interval and σ a positive number, there exists I_1, a subinterval of I_0 which is free from points of S_σ. Similarly there exists I_2, a subinterval of I_1 which is free from points of $S_{\sigma/2}$, and so on. In this manner a nested sequence of intervals I_n is obtained with the property that I_n is free from points of $S_{\sigma/2^{n-1}}$. Since $\bigcap_{n=1}^{\infty} I_n \neq \emptyset$, it follows that I_0 contains points of continuity of the function, and the theorem is proved. (The intervals I_n should, of course, be closed; Hankel slurred over this detail.) Again, this topological characterization of the points of continuity of an integrable function served to obscure the measure-theoretic property of this set of points.

It should be noted that in the two examples cited above Hankel employs the phrase "total magnitude s" in a slightly different sense than Riemann had. With Riemann it had always been used in the context of a partition of the fundamental interval $[a,b]$; it referred to the sum of those intervals of the partition in which the oscillation of the function is greater than σ. Hankel used the term to refer to intervals not directly related to a partition and which may be infinite in number. In fact, contained in Hankel's method for showing that the total magnitude of the intervals containing the points $1/2^n$ can be made arbitrarily small is the idea behind the proof that any countable set has Lebesgue measure zero.

All this might seem to suggest an anticipation of Borel and Lebesgue. But it must be kept in mind that the sets that Hankel was covering were of a very simple nature—so simple that each point of the set could be enclosed in an interval which would not intersect the corresponding intervals about the other points. Thus, the enclosing intervals corresponded to an infinite partition of the interval which, in turn, could be directly related to finite partitions and Riemann's condition (R_2). In his example, Hankel could just as well have enclosed S_σ in a finite number of intervals by employing the interval $[0,\epsilon]$ which includes all but a finite number of points of S_σ, and then covering the remainder as before. As long as the sets to be covered are the S_σ and the context is Riemann's theory of integration, it is, in fact, more natural to continue to use a finite number of intervals because they can more easily be related to partitions and thus to (R_2). It is therefore not surprising that Hankel's successors returned to the use of finite intervals and eventually developed a theory of measure based on finite covers. Furthermore, it must be recalled that the idea of a measure of a set of points was as foreign to Hankel's thinking as it was to Riemann's.

The most enthusiastic exponent of the new theory of integration was Paul du Bois-Reymond (1831–89), brother of the physiologist Emil du Bois-Reymond. During the early 1870's he became familiar with the work of Riemann, Hankel, Heine, and Cantor; by the fall of 1873 he had completed the manuscript for his "Versuch einer Classification der willkürlichen Functionen reeler Argumente nach ihren Aenderungen in den kleinsten Intervallen" [1875a].[7] This work documents his

---

7. Publication was held up until du Bois-Reymond agreed to make some minor deletions. See letters to him from Weierstrass [1923a: 199ff] on November 23 and 30 of 1873.

assimilation of the conceptual developments and indicates the perspective with which they were viewed in the 1870's.

In particular, du Bois-Reymond accepted Dirichlet's conception of a function as a correspondence. Functions in this general sense were referred to as "assumptionless functions" because, as Hankel had already stressed, they need not possess the properties formerly ascribed to all functions. According to du Bois-Reymond, the weakest restricting condition on a function is that it be integrable in Riemann's sense; integrable functions were thus regarded as forming the largest subclass of the class of assumptionless functions. This characterization reflects the manner in which Riemann's immediate successors regarded Riemann's condition for integrability. From the vantage point of the Lebesgue theory of the integral, it is true that Riemann's theory of integration appears insufficiently general. But in the 1870's—and even in the 1880's to a large extent—the class of integrable functions appeared to encompass a vastly larger domain of functions than had hitherto been conceived. Furthermore, the definition of the definite integral as the unique limiting value of Cauchy sums—the definition which served as the basis of Riemann's theory—was unquestioningly accepted as *the* natural manner of defining the integral. Riemann's extension was therefore regarded as the ultimate. It seemed, as du Bois-Reymond [1883a: 274] put it, that Riemann had extended the concept of an integrable function to its outermost limits.

Riemann's example of an integrable function whose points of discontinuity form a dense set showed that Dirichlet's integrability condition (the points of discontinuity must form a nowhere dense set) is not necessary for Riemann integrability. The condition is also not sufficient: Let S denote a nowhere dense subset of [0,1] which has positive content. Then $f(x) = \chi_S(x)$ satisfies Dirichlet's condition but fails to be integrable, since its oscillation is 1 at each point of S. Hankel thought that sets such as S did not exist. That appears to have been the opinion of du Bois-Reymond as well, for he regarded Dirichlet's condition as more restrictive than Riemann's, that is, as a sufficient condition for integrability [1875a: 22].

The reason for this mistake is fairly clear. By the time du Bois-Reymond wrote his paper, Cantor had introduced the class of first species sets, that is, sets P such that $P^{(n)}$ is a finite set for some integer n. These sets are, of course, nowhere dense and are certainly more complicated than anything imagined by Hankel. But it is not difficult to establish the fact that these sets can be enclosed in intervals of arbitrarily small total length. Thus, if first species sets are mistakenly assumed to exhaust the possibilities of nowhere dense sets, Hankel's false proposition will appear to be valid and Dirichlet's condition to be a special case of Riemann's.

This is probably what du Bois-Reymond thought, because he apparently did equate first species sets and nowhere dense sets. The following passage suggests this and is worth quoting in its entirety.

Points which do not form any lines can occur on an interval in two ways when infinite in number. *First,* so that in every arbitrarily small segment such points exist, as in the distribution of the rational numbers. *Secondly,* in every arbitrarily small part of the interval a finite segment can be specified on which none of these points lie.

In the second case, which we wish to consider more closely, the points are either finite in number or they become infinitely dense only upon approaching individual points. For if they are infinite in number, all of their distances cannot be finite. But also all of their distances in an arbitrarily small segment cannot vanish since otherwise the first case above would obtain. Thus their distances can become zero only at points or, more correctly, infinitely small segments.

The following distinctions have to be made here. 1. The points k_1 (as we shall denote them) become infinitely dense only upon approaching a finite number of points k_2. 2. The points k_2 themselves become infinitely dense upon approaching individual points k_3. 3. The points k_3, in turn, become infinitely dense upon approaching points k_4, etc.

Thus the roots of $0 = \sin(1/x)$ become infinitely dense in the vicinity of $x = 0$. Those of $0 = \sin [1/\sin(1/x)]$ in the vicinity of the roots of $\sin(1/x)$, etc., so that the distribution of the roots of

$$0 = \sin \cfrac{1}{\sin \cfrac{1}{\sin \cfrac{\cdot^{\cdot^{\cdot}}}{\cfrac{1}{\sin \cfrac{1}{x}}}}}$$

affords an illustration of the described distribution of points. [du Bois-Reymond 1875a: 35–36.]

The roots of $\sin(1/x) = 0$, $\sin[1/\sin(1/x)] = 0$, $\cdots$ form sets of types 1, 2, $\cdots$. Apparently du Bois-Reymond thought that if an infinite set of points becomes "infinitely dense" only at points and not on entire intervals, then these points can be classified according to the hierarchy characteristic of first species sets. And in a subsequent paper [1879] he implied that if a set is not of the first species then it must be dense in some interval (See 3.4). Such an interpretation of the structure of nowhere dense sets represents a continuation of the tendency noted in Lipschitz' paper.

Thus, du Bois-Reymond confounded the concept of a nowhere dense set with that of a first species set; as a result he would not have recognized any distinction between these notions and that of a set which is enclosable in a finite union of intervals of arbitrarily small total length. And the lack of counterexamples to, or even specific criticisms of, Hankel's proposition strongly suggests that the sort of confusion exhibited by du Bois-Reymond was widespread.[8]

Indeed, the first treatise on the theory of functions of a real variable, the *Fondamenti per la teorica delle funzioni di variabili reali* [1878] of Ulisse Dini (1845–1918), accorded an undue prominence to Cantor's first species sets. Dini was the first to state and prove that first species sets can always be enclosed in a finite number of intervals of arbitrarily small total length [1878: Sec. 14]. Using this property, he also obtained many other results which illustrate the negligible character of these sets with respect to the existence and value of the integral. For example, if the discontinuities of a bounded function form a first species set, then the function is integrable; or if f is integrable and $f(x) = g(x)$ (g bounded) except on a first species set, then g is likewise integrable and has the same integral as f [1878: Sec. 187ff]. For the vast majority of such theorems, the same proofs clearly go through unaltered when first species sets are replaced by any set that can be enclosed in a finite number of intervals of arbitrarily small total length—that is, a set of zero content—because precisely this property of first species sets is used. But Dini failed to extricate it from the context of first species sets. It failed to suggest to him the value of a theory of content or even a theory of sets of content zero.

It is doubtful that Dini actually identified first species sets with nowhere dense sets, because he did express doubt about the general validity of Hankel's proposition. Since he knew that first species sets have content zero, their identification with nowhere dense sets would

---

8. Hankel's "Untersuchungen" [1870] was reviewed by Cantor [1871b] and by Jules Houël [1870]. Neither of them was critical of Hankel's proposition.

have led him to accept the proposition. Nevertheless, Dini's preoccupation with first species sets and his inability to make a substantial criticism of Hankel reflect his own limited understanding of the actual differences between these concepts and that of a set of zero content. This point will be further substantiated in 2.3.

What is perhaps the most striking and unfortunate example of the confusion of these concepts occurs in a paper [1880] on trigonometric series by Axel Harnack (1851–1888), brother of the theologian Adolf Harnack. The importance of first species sets, according to Harnack, is due to the fact that "they exert no influence on the value of the integral . . . " [1880: 126]. Thus he defines a set to be of the first species if it can be enclosed in a finite number of intervals of arbitrarily small total length. In so defining these sets, Harnack did not think he was doing anything more than reformulating Cantor's original definition. He assumed that the property described in his definition implies that a set has the structure of Cantor's first species sets, for he made crucial use of results that Cantor had obtained by induction on the type of the first species set—a mistake that vitiated Harnack's results.

As early as 1875, however, H. J. S. Smith (1826–83), Savilian Professor of Geometry at Oxford, published a paper "On the Integration of Discontinuous Functions" [1875] which, had it been known to mathematicians on the Continent, would have helped clarify these notions.

Smith's principal work was in the theory of numbers. But he had studied in Paris, had spent a good deal of time living on the Continent, and was familiar with the French, German, and Italian languages. As a mathematician his inspiration came from the Continent, and he kept abreast of the latest developments. In particular he read the papers of Riemann and Hankel and came to question, among other things, Hankel's assertion that a pointwise discontinuous function is always integrable. Smith discovered two entirely different methods for constructing nowhere dense sets—sets "in loose order" in his terminology —and the second provided him with a counterexample to Hankel's assertion.

The first method of construction represents the logical continuation of the method adumbrated by Hankel's sole example of a nowhere dense set—the set of points of the form $1/2^n$. Smith started with the analogous set, P_1, which consists of all points of the form $1/n$, where n denotes a positive integer, and proceeded by induction to generalize P_1: Let P_2 denote the set of points of the form $(1/n_1)+(1/n_2)$, and, more generally, let P_s denote the set of points of the form $(1/n_1)+ (1/n_2)+ \cdots +(1/n_s)$, where $n_1, n_2, \cdots, n_s$ are positive integers.

Then the points of P_2 are "indefinitely condensed" in the vicinity of the points of P_1, although P_2 is still nowhere dense. Smith gave a neat inductive proof that P_s is nowhere dense for all values of s. Cantor's definition of first species sets was not known to Smith. In terms of Cantor's notions, however, P_s is of the sth type, since:

$$P_s' = \{0\} \cup P_1 \cup P_2 \cup \cdots \cup P_{s-1};$$

$$P_s{}^{(n)} = \{0\} \cup P_1 \cup \cdots \cup P_{s-n};$$

$$P_s{}^{(s)} = \{0\}.$$

Smith's second method was altogether different:

> Let m be any given [integer] greater than 2. Divide the interval from 0 to 1 into m equal parts; and exempt the last segment from any subsequent division. Divide each of the remaining $m-1$ segments into m equal parts; and exempt the last segment of each from any subsequent division. If this operation be continued *ad infinitum*, we shall obtain an infinite number of points of division P upon the line from 0 to 1. These points lie in loose order . . . . [1875: 147.]

The set P together with its limit points forms what is nowadays called a Cantor set. By virtue of the method of construction, it is easily seen that P is nowhere dense. Furthermore, after k operations in the construction defining P, the total length of the nonexempted intervals is $(1-1/m)^k$ "so that as k increases without limit, the points of division of P occur upon segments which occupy only an infinitesimal portion of the interval from 0 to 1" [1875: 147].

A bounded function whose points of discontinuity form one of the sets P_s or the set P will therefore be integrable because these sets can be covered by intervals of arbitrarily small total length. But that property is not shared by all nowhere dense sets, as Smith was able to show by slightly modifying the construction used to define P. The idea is to "divide the interval from 0 to 1 into m equal parts, exempting the last segment from any further division; . . . divide each of the . . . remaining $m-1$ segments by m^2, exempting the last segment of each segment; . . . again divide each of the remaining $(m-1)(m^2-1)$ segments by m^3, exempting the last segment of each segment, and so on continually" [1875: 148]. In other words, the idea is to diminish the length of the deleted segments more rapidly than in the construction of P.

The resulting set Q then has the property that at the kth stage in its construction, the length of the nonexempted segments is $(1-1/m)(1-1/m^2) \cdots (1-1/m^k)$ rather than $(1-1/m)^k$; and the former approaches the positive limit $\prod_{k=1}^{\infty} (1-1/m^k)$ as $k \to \infty$. (Consequently, $c_e(Q) = \prod_{k=1}^{\infty} (1-1/m^k)$.) From this it follows easily that there exist bounded, pointwise discontinuous functions which are not integrable, e.g., $f(x) = \chi_Q(x)$.

Smith also made an interesting observation in connection with Hankel's proof that a nowhere dense set can be enclosed in intervals of arbitrarily small total length. The proof, he felt, "ceases to convey any clear image to the mind, as soon as the number of points becomes infinite." Suppose, for example, the nowhere dense set in question is P_{s+1}. Then if [0,1] is divided in such a manner that each point of P_{s+1} is in a segment of its own, the lengths of these segments in the neighborhood of the points of P_s will have no lower bound, "and if such a mode of division is admissible, it is difficult to see why it should not also be admissible so to divide the line as to include every rational point in a segment of its own: in which case Dr. Hankel's proposition would extend to systems in close order . . . " [1875: 149–50].

What Smith seems to be saying here is that if an infinite number of intervals are allowed in order to enclose the points of a set, then the dense set of rational numbers can be covered by intervals of arbitrarily small total length. He may have realized that Hankel's method of enclosing the points $1/2^n$ in an infinite number of intervals of arbitrarily small total length can be extended to the set of rational numbers, since they can be set in one-to-one correspondence with the positive integers.[9] But if a function makes jumps greater than some σ at the rational points, it cannot be integrable. This criticism of Hankel's careless use of an infinite number of intervals is thus perfectly correct. At the same time, however, Riemann's integrability condition (R₂) is equivalent to the requirement that the set of points of discontinuity, rather than the sets S_σ, be enclosable in an infinite number of intervals of arbitrarily small total length. The idea of a theory of measure, in contrast to the theory of content Hankel had vaguely suggested, would undoubtedly have been developed much earlier had its relevance to Riemann inte-

9. Cantor [1874] introduced the concept of countability and proved that the algebraic numbers form a countable set; of course, Smith could easily have realized the countability of the rational numbers independently of Cantor's work.

grability been known. (The above-mentioned equivalence was discovered independently by Lebesgue, G. Vitali, and W. H. Young in the early 1900's. For Vitali and Young it was one of the factors that induced them to develop the equivalent of Lebesgue's theory of measure, independently of each other and of Lebesgue. See 5.2.)

Probably the development of a measure-theoretic viewpoint within integration theory would have been accelerated had the contents of Smith's paper been known to mathematicians whose interest in the theory was less tangential than Smith's. Unfortunately the reviewer for *Fortschritte der Mathematik* failed to grasp the theoretical significance of Smith's examples and simply noted that the paper contained a more precise formulation of Riemann's integrability condition and a more rigorous proof.[10] As a result, it was not until after Continental mathematicians in the early 1880's discovered similar examples of nowhere dense sets with positive outer content that Smith's work became known to them.

In the same paper Smith also disproved Hankel's theorem that when a bounded function f is integrable the one-sided limits $f(x-)$ and $f(x+)$ must exist. It was also disproved by the German mathematician Karl J. Thomae (1840–1921). Thomae's *Einleitung in die Theorie der bestimmten Integrale* [1875] is devoted to a rigorous development of the properties of the Riemann integral. Here Thomae noted that the function $f(x) = \sin[1/\sin(\pi/x)]$ is integrable although $f(x-)$ and $f(x+)$ fail to exist for $x = 1/m$, for $m = 1, 2, 3, \cdots$ [1875: 15–16]. Thomae, however, failed to find fault with Hankel's identification of pointwise discontinuous and integrable functions.

When Riemann gave his integrability criterion (R_1), he claimed that it represented the necessary and sufficient condition for the Cauchy sums of a function to converge to a unique limit (in the sense that, corresponding to every $\epsilon > 0$, a $\delta > 0$ can be determined so that all Cauchy sums deriving from partitions of norm less than δ differ by less than ϵ from a fixed limiting value). It is not immediately clear, however, that this follows from (R_1). To establish this point, Thomae introduced the so-called upper and lower Riemann sums,

$$U(P) = \sum M_i \delta_i \quad \text{and} \quad L(P) = \sum m_i \delta_i,$$

corresponding to a partition P of $[a,b]$ into intervals of lengths δ_i. He discovered that these sums always converge to unique limits as $\|P\| \to 0$, whether or not the bounded function under consideration is integrable. This result was also published in 1875 independently by

10. *Fortschr. Math., 1875* (1877), 247. The reviewer was F. Müller of Berlin.

Giulio Ascoli [1875], H. J. S. Smith [1875: 141ff], and Gaston Darboux [1875: 64ff]. It was not until 1881, however, that the terms "upper integral" and "lower integral" were introduced by Vito Volterra [1881b: 338ff] along with the suggestive notation $\underline{\int_a^b} f(x)dx$ and $\overline{\int_a^b} f(x)dx$ for the limiting values of $L(P)$ and $U(P)$, respectively. Riemann's (R_1) is thus transformed into the requirement that $\underline{\int_a^b} f(x)dx = \overline{\int_a^b} f(x)dx$.

This form of the criterion motivated Giuseppe Peano to make the distinction between inner and outer content and to introduce the requirement of measurability in analogy with (R_1). (See 4.1.)

Independently of Thomae, du Bois-Reymond had also come to realize that associated with every bounded "assumptionless" function are two continuous functions, $G(x) = \underline{\int_a^x} f$ and $K(x) = \overline{\int_a^x} f$. These functions fascinated him, and in a review of Thomae's *Einleitung* [du Bois-Reymond 1875c], he indicated that he had occupied himself considerably with them and, specifically, with the question: When f is an "assumptionless function," for what integrable functions g and k is it true that $G(x) = \int_a^x g$ and $K(x) = \int_a^x k$? The phrasing of the problem suggests that du Bois-Reymond was optimistic about the existence of g and k. In fact, he promised a future paper on this matter although it never materialized. That may have been caused by the fact that he soon came to realize that the existence of *integrable* g and k cannot be established in general. It will be instructive to speculate on the manner in which he could have realized this.

In his researches on trigonometric series, du Bois-Reymond found it necessary to apply the formula for integration by parts in the context of Riemann-integrable functions. His justification of the formula is based upon the following idea [1875b: 129–33]: Let P denote a partition of $[a,b]$ and consider the function

$$f_P(x) = \sum m_j \chi_{I_j}(x),$$

where the I_j denote the intervals into which P divides $[a,b]$ and the m_j denote as usual the greatest lower bound of f on I_j. Then the validity of integration by parts can be extended to integrable functions by the facts that $\int_a^b f_P$ converges to $\int_a^b f$ as $\|P\| \to 0$ and that integration by parts is obviously valid for functions such as the f_P. Incidentally,

Lebesgue, who admired du Bois-Reymond's work on trigonometric series, later made effective use of this technique of proof in connection with his own theory of integration (See 5.1).

It is not difficult to see now that the natural candidate for du Bois-Reymond's desired function g satisfying $G(x) = \int_a^x g$ would be $g(x) = \lim_{n \to \infty} g_n(x)$, where $g_n(x) = g_{P_n}(x)$ and P_n denotes a sequence of partitions such that P_{n+1} is a refinement of P_n and $\|P_n\| \to 0$. The function g is well defined since the sequence $g_n(x)$ is decreasing for each value of x. But g must still meet two conditions: g must be Riemann-integrable and $\overline{\int_a^x} g = \underline{\int_a^x} f$, that is, $\int_a^x g = \lim_{n \to \infty} \int_a^x g_n$. Here du Bois-Reymond would have been face to face with the intrinsic defects of Riemann's definition, since these conditions cannot be affirmed in general. (They are, of course, characteristic properties of the Legesgue integral.) It is therefore not surprising that his promise of a paper on this subject was never fulfilled.

Underlying du Bois-Reymond's problem, of course, is the question: Under what conditions is a continuous function $G(x)$ an integral, that is, expressible in the form $G(x) = \int_a^x g$? This question was later raised by Harnack in another context (3.3), but Lebesgue was the first to provide a satisfactory answer (5.1).

2.3   DISCUSSIONS OF THE DIFFERENTIABILITY PROPERTIES OF CONTINUOUS FUNCTIONS

This section serves two purposes. Its immediate objective is an analysis of Dini's work on the four derivatives of a continuous function and the developments that motivated it. At the same time it provides the background essential for an appreciation of Lebesgue's important discovery that continuous functions of bounded variation are differentiable except on a set of measure zero. Functions of bounded variation are defined in 3.4. For an appreciation of the developments in the present section in relation to Lebesgue's discovery, it is important to note that these functions can be defined equivalently as those functions which are representable as the difference of two increasing functions. Hence Lebesgue's theorem implies, among other things, that a continuous function with at most a finite number of maxima and minima is differentiable except possibly on a set of measure zero.

Discussions of the existence of the derivative of an arbitrary function (explicitly assumed to be continuous by later authors) have their origin in the reformation of the foundations of the calculus proposed by Joseph Louis Lagrange (1736–1813). He suggested [1772] that the differential calculus be based on the fact that if f is any function, then

$$(4) \qquad f(x + i) = f(x) + pi + qi^2 + \cdots,$$

where i is a real number and $p, q, \cdots$ are functions of x alone. From (4) he deduced by purely formalistic reasoning that $f'(x) = p$, $f''(x) = 2q$, etc. The advantage of this approach was that it avoided the metaphysical difficulties that arise with the use of infinitely small or evanescent quantities—difficulties which were very real in the eighteenth century—and he subsequently made this approach the basis of his *Théorie des fonctions analytiques . . . contenant les principes du calcul différentiel . . . réduites à l'analyse algébrique des quantités finies* [1797].

In the *Théorie* Lagrange proposed to give the first a priori demonstration that (4) holds except for certain "particular" values of x (assumed finite in number). The argument is based upon the assumption that for any fixed value of x,

$$(5) \qquad f(x + i) = \sum A_k i^{r_k},$$

where the r_k are rational numbers and the A_k are constants [1797: 7–8]. Lagrange never questioned (5). His proof aims at showing the r_k must be positive integers so that (5) reduces to (4). As Lagrange [1803: 9–10] later admitted, his proof is only meaningful when the function f is assumed to be algebraic, but he shrugged off this point. Many of Lagrange's contemporaries, however, were critical of his conclusions or at least his proof. (See [Dickstein 1899].)

An immediate corollary to (4) is the proposition that any function possesses a derivative in general, that is, except possibly at the points where (4) fails to hold. A. M. Ampère [1806] sought to establish this corollary independently of (4) and then use it to establish (4) without recourse to Lagrange's argument. It is his proof that became the paradigm for subsequent ones. What Ampère actually showed is that if $f'(x)$ exists for all x ($f'(x)$ can be infinite for some values of x) and if the difference quotient converges uniformly to $f'(x)$, then any interval (a,k) must contain a point x_0 at which $f'(x_0)$ is finite and nonzero. This would imply that $f'(x)$ exists as a finite, nonzero quantity on a dense set. But Ampère interpreted his result to mean that $f'(x)$ is neither zero nor infinite except for "certain particular and isolated values of x" [1806: 149].

Between the appearance of Ampère's paper and 1870, the proposition that any (continuous) function is differentiable in general was stated and proved in most of the leading texts on the calculus.[11] Frequently Ampère's original argument was simplified by making use of the "intuitively evident" fact that a function which varies continuously must be piecewise monotonic. Thus, differentiability and monotonicity were linked together, albeit tenuously. This is particularly interesting because Cauchy in his *Leçons sur le calcul différentiel* [1829] had considered the functions $f(x) = x^3 \sin(1/x)$ and $g(x) = x \sin(1/x)$, both of which have removable discontinuities at $x = 0$ and yet are not monotonic in any neighborhood of 0. The function g, of course, also contradicts the assumption that $\dfrac{g(x + h) - g(x)}{h}$, the difference quotient, always approaches a fixed limiting value. And, although Cauchy did not call attention to this point, he was fully aware that if $\lim_{h \to 0} F(h) = \lim_{h \to 0} G(h) = 0$, it can happen that $F(h)/G(h)$ does not approach a unique limiting value as h approaches zero. (See [Cauchy 1829: 331, 337–38].)

The most detailed attempt to establish the differentiability of a continuous function was made by the Belgian engineer and mathematician A. H. E. Lamarle (1806–75). He recognized [1855] the need to demonstrate the assumptions that a continuous function is piecewise monotonic and that the difference quotient does not oscillate indefinitely, and he provided "proofs" of these points. Lamarle believed he had established that the difference quotient of a continuous function could not converge to zero, become infinite, or fail to approach a unique limiting value on a dense set of points. This would imply that the points at which a continuous function fails to possess a finite, nonzero derivative must form a nowhere dense set. But Lamarle concluded that these points "remain isolated from one another, and maintain a certain interval between them" [1855: 40–41], thereby indicating that he also failed to comprehend the possibilities of a nowhere dense set. Lamarle's lack of comprehension—and Ampère's as well—illustrates the same type of faulty reasoning with infinite sets that served to obscure the significance of measure-theoretic properties of sets in connection with Riemann's (R_2) and further indicates how easy it was to go astray in these matters.

By extending the integral concept to discontinuous functions, Riemann had actually revealed the untenability of these proofs. For, as Hankel [1871: 199–200] pointed out, the function

---

11.   Lacroix 1810–19: I, 241n; Raabe 1839–47: I, 7ff; Duhamel 1847: I, 22; 1856: I, 94–97; Freycinet 1860: 39–42; Bertrand 1864–70: I, 2–4; Serret 1868: I, 16–21; Rubini 1868: I, 36–37. Two proofs were also attempted by Galois [1962: 383, 413ff].

$$F(x) = \int_0^x \sum_{n=1}^{\infty} (nx)/n^2$$

is everywhere continuous, but $F'(x)$ does not exist on the dense set of points at which $\sum_{n=1}^{\infty} (nx)/n^2$ is discontinuous. By 1861 both Riemann and Weierstrass were pointing out in their lectures that differentiability is not a consequence of continuity.[12] Hankel was the first, however, to publish a counterexample. He also observed that the idea behind Riemann's example can be generalized to a method of constructing functions that have a singularity on the dense set of rational numbers from functions that have a singularity at a single point. This method he called the principle of condensation of singularities [Hankel 1870: 77ff]. For example, if g is continuous on $[-1,1]$ but $g'(0)$ fails to exist, then, according to Hankel, the function

$$f(x) = \sum_{n=1}^{\infty} g(\sin n\pi x)/n^s$$

is continuous but not differentiable at rational points for $s > 2$.

Hankel's arguments were neither complete nor entirely correct, and they evoked a critical reply from the Belgian mathematician Phillipe Gilbert (1832–92), who pointed out these weaknesses "so as to leave no doubt . . . about the inanity of the conclusions" [1873a: v]. Gilbert was convinced that continuous functions must be generally differentiable and that Lamarle's proof was basically sound. A large portion of his paper is devoted to presenting this proof in a slightly revised form.

One of Weierstrass' students, H. A. Schwarz (1843–1921), soon communicated to Gilbert a legitimate counterexample, defined as follows [1873]: Let $g(x) = [x] + (x-[x])^{1/2}$, where $[x]$ denotes the greatest integer less than or equal to x. (See Figure 2.) The counterexample is then $f(x) = \sum_{n=0}^{\infty} g(2^n x)/2^{2n}$, which can be shown to be nondifferentiable on the dense set of points of the form $m/2^n$, even though $f(x)$ is a strictly increasing function. Gilbert [1873b] responded by admitting that his conclusions had to be wrong, but he failed to grasp the extent to which his reasoning was defective and believed that he still had established the impossibility of a *nowhere* differentiable continuous function.

12. See [Weierstrass 1923a: 199–201] and [H. A. Schwarz 1873]. Examples of continuous, nowhere differentiable functions had been constructed—but not published—prior to 1861 by B. Bolzano [1930: 66–70, 98–99] and C. Cellérier [1890], but they had no influence on the developments under consideration.

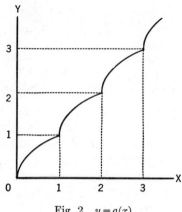

Fig. 2 $y = g(x)$

When Darboux composed his memoir on discontinuous functions, he was familiar with Hankel's "Untersuchungen" and the remarks of Gilbert and Schwarz; Darboux devoted a portion of his memoir to a study of the questions involved. In particular he was able to remove all doubts about the existence of nowhere differentiable continuous functions by showing that $f(x) = \sum_{n=1}^{\infty} \dfrac{\sin(n+1)!x}{n!}$ is such a function [1875: 107–08].[13] Another example of this type had already been constructed by Weierstrass [1894–1927: II, 71–74] and presented before the Berlin Academy on July 18, 1872, but it was first published by Paul du Bois-Reymond [1875a: 29–31].

Weierstrass also privately communicated another example to du Bois-Reymond to show that a monotonic or piecewise monotonic function need not be differentiable [Weierstrass 1923a: 206; see also 1923b: 231–32]. This appeared particularly necessary since one of his former students, Leo Königsberger, still believed it was possible to reconcile the new discoveries with the old belief that continuity entails differentiability at most points. Indeed, in a treatise on elliptic functions, Königsberger [1874: 13] posited the following proposition as the foundation of the theory of infinitely small and large quantities: If $f(x)$ is any nonconstant function with a finite number of maxima and minima and $f(x)$ does not possess finite discontinuity jumps on a dense set, then $f'(x)$ exists as a finite nonzero quantity except at particular points x. The last condition on f was clearly intended to rule out functions such as Riemann's $f(x) = \sum_{n=1}^{\infty} (nx)/n^2$. Schwarz's example suffices to show that

---

13.   Darboux [1879] later published a more detailed and general discussion.

Königsberger's proposition is untenable, but Weierstrass preferred his own example: Let $g(x) = \frac{1}{2}x\sin[\frac{1}{2}\log(x^2)]$. Then g is differentiable except at 0, where $\dfrac{g(h) - g(0)}{h} = 1 + \frac{1}{2}\sin[\frac{1}{2}\log(h^2)]$. If $0 < k < 1$ and a_1, $a_2, a_3, \cdots$ denotes a sequence of real numbers, then $f(x) = \sum_{n=1}^{\infty} k^n g(x - a_n)$ is continuous and monotonic but is not differentiable at $x = a_n$ for $n = 1, 2, 3, \cdots$. (At $x = a_n$, $D^+f(x) - D_+f(x) = D^-f(x) - D_-f(x) = k^n$, where $D^+f(x)$, etc., denote the Dini derivatives of f at x; see below in the discussion of Dini's work for the definitions.) Weierstrass pointed out that the set $\{a_n\}$ can be taken as the dense set of all algebraic numbers, since Cantor had proved that this set is countable. As for the possibility of totally nondifferentiable monotonic functions, Weierstrass confessed that it appeared difficult to construct one for which the difference quotient $\dfrac{f(x+h) - f(x)}{h}$ remains bounded, but he was convinced that such functions actually exist. Lebesgue's theorem referred to at the beginning of the section, of course, shows that they cannot exist.

Discussions of the differentiability of continuous functions were followed with great interest by Ulisse Dini (1845–1918), who had become a professor of higher analysis at the University of Pisa in 1871. Dini had been a student at Pisa and had spent the year 1865 on a scholarship in Paris. Upon returning to Italy, he began to have doubts about the proofs and statements of some of the fundamental propositions of analysis. In particular he must have objected to the proposition that continuous functions are differentiable, a proposition which was proved in all of the leading textbooks. These doubts were confirmed by the publication of the counterexamples of Hankel, Schwarz, Darboux, and Weierstrass. Weierstrass' example especially impressed him; Dini showed [1877a, 1877b] that this function is a special case of an entire class of similar functions defined by infinite series with terms which are differentiable functions.

In a subsequent paper that year, Dini [1877c] began by remarking that since the existence of nowhere differentiable functions had now been established beyond doubt, either limiting conditions must be imposed on the function concept to insure the applicability of the differential calculus or one must find "more general methods of the calculus which apply to any continuous function" [1877c: 131]. Dini chose the latter course. His results were incorporated into his widely read *Fondamenti per la teorica delle funzioni di variabili reali*, which appeared in 1878.

Dini [1878: Sec. 130 ff] began by introducing the numbers l and L, which represent the greatest lower bound and the least upper bound, respectively, of $\dfrac{f(x+h)-f(x)}{h}$ for x and $x+h$ in an interval $[a,b]$ and h positive. He observed that if either l or L is finite, then there exists a linear function $Ax+B$ such that $g(x)=f(x)+Ax+B$ is a strictly monotonic function on the interval $[a,b]$. For example, if l is finite and A is any number greater than $-l$, then

$$\frac{g(x+h)-g(x)}{h} = \frac{f(x+h)-f(x)}{h} + A > l + A > 0;$$

this shows that g is a strictly increasing function. Functions with the property that either l or L is finite were termed functions of the first species, and Dini observed that these functions can always be represented as the difference of two increasing functions. For example, if l is finite and A is taken greater than both $-l$ and 0, then $h(x)=Ax+B$ and $g(x)=f(x)+h(x)$ are strictly increasing functions and $f=g-h$. Dini's first species functions thus form a subclass of the class of functions of bounded variation (later introduced by Camille Jordan), for which such a representation is characteristic.

Thus a function f of the first species can be transformed into a monotonic function g by the addition of a linear function. On the other hand, g possesses the same differentiability properties as f. From this Dini concluded that, for first species functions at least, the existence of infinitely many maxima and minima or intervals on which the function is constant has nothing to do with the nonexistence of the derivative. This observation had a crucial and unfortunate influence upon Dini's work, for it led him to believe that if propositions such as Ampère's ($f'(x)$ exists on a dense set) are to be rigorously established, a condition stronger than that of monotonicity or "piecewise monotonicity" would be necessary—for example, the requirement that the functions $f(x)+Ax+B$ also be piecewise monotonic except for a finite number of values of A.

Dini's method of attack centered on the study of the functions

$$g(x;p,q) = f(x) - f(p) - \frac{f(q)-f(p)}{q-p}(x-p),$$

where $[p,q]$ denotes a subinterval of $[a,b]$, the interval on which f is assumed to be defined. The usefulness of these functions had been demonstrated by Ossian Bonnet in his proof of the mean value theorem, which has since become the standard proof [see Serret 1868: I, 17–19].

Similar functions had been introduced by Cantor [1870b].[14] Since $g(x;p,q)$ takes the value 0 at p and q, it possesses at least one relative maximum or minimum in the open interval (p,q). Translated into terms of f, this yields inequalities involving $\dfrac{f(x+h)-f(x)}{h}$ from which Dini was able to deduce some theorems concerning the differentiability properties of a certain class of continuous functions. For example, he established a rigorous version of Ampère's "theorem": If a continuous function f has the property that $f(x)+Ax+B$ possesses a finite number of maxima and minima (so that it is piecewise monotonic) for all values of A with the exception of a finite number, then $f'(x)$ exists on a dense set [Dini 1878: Sec. 143]. (See the Appendix for a sketch of Dini's interesting proof.) Dini's hypothesis on the function f is so strong that he was able to derive other information about its differentiability properties such as that f possesses one-sided derivatives at all points (possibly with infinite values at some) [1878: Sec. 162]. Monotonicity and differentiability thus appeared to be closely linked as a result of Dini's work, but it still left open the question as to whether continuous monotonic functions can exist which are nowhere differentiable. Certainly a reader of *Fondamenti* would come away thinking the answer is likely to be affirmative.

One of Dini's most significant contributions to the study of continuous functions are the four derivatives of a continuous function f, $D_-f(x)$, $D^-f(x)$, $D_+f(x)$, and $D^+f(x)$, which he introduced as follows [1878: Sec. 145].[15] For x in (a,b) let l_x and L_x denote the greatest lower bound and the least upper bound of $\dfrac{f(x+h)-f(x)}{h}$ for h satisfying $0<h<b-x$. Then l_x and L_x are functions of x and b. As b decreases to x, l_x increases and L_x decreases. Consequently,

$$D_+f(x) = \lim_{b \searrow x} l_x \quad \text{and} \quad D^+f(x) = \lim_{b \searrow x} L_x$$

exist and depend on x alone. $D_-f(x)$ and $D^-f(x)$ are defined similarly with respect to $\dfrac{f(x-h)-f(x)}{-h}$ for $h>0$. In more modern notation,

$$D_+f(x) = \liminf_{h \searrow 0} \frac{f(x+h) - f(x)}{h} \; ; \quad D^+f(x) = \limsup_{h \searrow 0} \frac{f(x+h) - f(x)}{h} \; .$$

It is interesting that the similar functions

$$\liminf_{h \to 0} \frac{f(x+h) - f(x)}{h} \quad \text{and} \quad \limsup_{h \to 0} \frac{f(x+h) - f(x)}{h}$$

had already been introduced by Lamarle and, with greater perspicuity, by Gilbert [1873a: 26] in order to show that $f'(x)$ must exist! It is possible that Dini was influenced by Gilbert's paper, since he knew of its existence. The fundamental property of the four derivatives that Dini discovered, through his study of the functions $g(x;p,q)$, is that if λ and Λ denote the greatest lower bound and the least upper bound of any one of these derivatives on some interval I, then $\lambda = l$ and $\Lambda = L$. Thus, all four derivatives have the same greatest lower bound and the same least upper bound on any interval. It should be noted that this implies that whenever one of the four derivatives is integrable, so are the rest, and that they all possess the same integral over any interval.

The four derivatives provided Dini with an analytical tool suitable for a more general and elegant study of the Fundamental Theorem in the context of integrable functions [1878: Sec. 197ff]. In Cauchy's version of this theorem (1.2), that is, in the context of continuous functions, the Fundamental Theorem I states that, for all x,

(6)
$$\left(\int_a^x f \right)' = f(x).$$

Cauchy had used (6) to establish the Fundamental Theorem II, viz.,

(7)
$$\int_a^b F' = F(b) - F(a),$$

where F is any function with a continuous derivative F'. But, as Hankel had pointed out with Riemann's example $f(x) = \sum_{n=1}^{\infty} (nx)/n^2$, (6) can fail to hold for values of x forming a dense set when f is assumed simply to be integrable. (If $F(x) = \int_a^x f$ and f denotes Riemann's function, then at every x, F has one-sided derivatives which are equal to $f(x-)$ and $f(x+)$.)

Darboux [1875: 111–12][16] showed that the mean value theorem af-

16. Here Darboux made crucial use of Bonnet's version of the mean value theorem. Earlier versions such as Cauchy's had included the hypothesis that the function in question has a continuous derivative.

fords an elegant proof of (7) when F' is assumed integrable, since if P is a partition, $a = x_0 < x_1 < \cdots < x_n = b$, of $[a,b]$, then

$$(8) \quad F(b) - F(a) = \sum_{i=1}^{n} F(x_i) - F(x_{i-1}) = \sum_{i=1}^{n} F'(t_i)(x_i - x_{i-1}),$$

where t_i is in (x_{i-1}, x_i). The theorem follows because the right-hand side converges to $\int_a^b F'$ as $\|P\| \to 0$. Independently of Darboux, du Bois-Reymond had established (7) with F' replaced by one of the one-sided derivatives of F—under the hypothesis that it exists—but his proof was extremely long. The mean value theorem does not apply to one-sided derivatives, and consequently the method of (8) cannot be imitated directly.

Despite his effort "to secure the crumbling foundations of the integral calculus," du Bois-Reymond [1875b: 166] concluded his proof on a note of uncertainty; he had not shown the impossibility of another definition of a generalized derivative with the property that the generalized derivative of some continuous function F is integrable but the integral of the derivative over $[a,b]$ differs from $F(b) - F(a)$. Du Bois-Reymond's concern with another definition of the derivative may have been due to his realization that $F(x) = \int_a^x f$ need not possess one-sided derivatives when f is integrable. The four derivatives enabled Dini to clear up this point, since if $F(x) = \int_a^x f$ then the four derivatives of F are bounded and, in fact, integrable; and $F(b) - F(a) = \int_a^b DF$, where DF denotes any one of Dini's derivatives.

Thus, the Fundamental Theorem I retains a meaning for integrable functions in the sense that $F(x) = \int_a^x f$ possesses four derivatives which differ from each other and from f by "functions of null integral" [Dini 1878: Sec. 202]. That is, $DF(x) = f(x) + n(x)$, where n is an integrable function whose integral over every interval is zero. Since an integrable function is a function of null integral if and only if it is zero except possibly on a set of measure zero, Dini's result is equivalent to the proposition that $\left(\int_a^x f\right)' = f(x)$ except on a set of zero measure. Measure-theoretic considerations such as these were, of course, foreign to Dini's thinking. But the introduction of the concept of functions of null integral reflects his realization that the relationship between DF and f

is not completely characterized by the fact that $DF(x) = f(x)$ on a dense set. Dini was also familiar with the example of a function of null integral that Thomae [1875: 14] had constructed: $f(p/q) = 1/q$ where p and q are relatively prime, and $f(x) = 0$ when x is irrational. This function is positive on a dense set and zero on a dense set.

Next Dini considered the Fundamental Theorem II and provided a proof as elegant as Darboux's and yet more general than du Bois-Reymond's. The fundamental property of the four derivatives, the fact that $\lambda = l$ and $\Lambda = L$, shows that

$$\frac{F(x_i) - F(x_{i-1})}{x_i - x_{i-1}} = D_i,$$

where D_i lies between the greatest lower bound and the least upper bound of DF on $[x_{i-1}, x_i]$; this equation enabled Dini to generalize (8). Thus if F is a continuous function and one of its four derivatives DF is integrable, $\int_a^b DF = F(b) - F(a)$.

In their versions of this theorem, Darboux and du Bois-Reymond had been careful to include the hypothesis that the derivative of F be integrable, but it was Dini [1878: Sec. 200] who first called attention to the fact that this hypothesis is not superfluous, even when the derivatives of F are bounded. Suppose, he reasoned, that F is not constant on $[\alpha, \beta]$ but that every subinterval of $[\alpha, \beta]$ contains points at which F takes on a relative maximum or minimum value or contains an interval in which F is constant. Suppose further that the four derivatives are bounded. Then $D_+F(x) \leq 0$ and $D^+F(x) \geq 0$ on dense subsets of $[\alpha, \beta]$ so that, corresponding to any partition $\alpha = x_0 < x_1 < \cdots < x_n = \beta$, there exist points t_i and t_i' in (x_{i-1}, x_i) such that $\sum_{i=1}^{n} D_+F(t_i)(x_i - x_{i-1}) \leq 0$ and $\sum_{i=1}^{n} D^+F(t_i')(x_i - x_{i-1}) \geq 0$. Thus, if the four derivatives of F were integrable, it would be necessary that $\int_\alpha^\beta DF = 0$; but the Fundamental Theorem II implies that $F(\beta) - F(\alpha) = \int_\alpha^\beta DF$. Since F is not constant on $[\alpha, \beta]$, however, it can be assumed without loss of generality that $F(\beta) - F(\alpha) \neq 0$. Hence the derivatives of F cannot be integrable.

Actually Dini had to admit that his conclusion was only "very likely," since he was unable to construct a continuous, nonconstant function F with the property that every interval contains a subinterval on which F remains constant. The existence of such a function is directly related to the existence of nowhere dense sets of positive outer content: Suppose, that F is a continuous function defined on [0,1] with densely distributed "intervals of invariability" $I_1, I_2, I_3, \cdots$. For simplicity these intervals will be taken as open intervals; that is, their endpoints are not included. Let G denote the set of points in [0,1] which do not belong to the I_n. Then G is a nowhere dense set by virtue of the ubiquity of the I_n. Furthermore, the points of discontinuity of, for example, D^+F must belong to G since D^+F is identically zero on each I_n and consequently continuous there. Therefore, since D^+F is not integrable, its points of discontinuity cannot be enclosed in a finite number of intervals of arbitrarily small total length. In other words, G must be a nowhere dense set of positive outer content. It is important to realize that these remarks also suggest the manner for constructing a nowhere dense set of positive outer content, namely by constructing a sequence of densely distributed intervals in, for example, [0,1] such that the sum of their lengths is less than 1. Then G, the complementary set of the intervals, has positive outer content.

Dini was confident that his hypothetical functions actually exist, and in fact he conjectured further that there exist functions which possess an ordinary, bounded derivative and which take on extremal values in every interval. Du Bois-Reymond had conjectured that, on the contrary, a function that is everywhere differentiable cannot have maxima occurring on a dense set [1875: 32]. On this point Dini was later proven correct, and his conjectures led to the discovery of entire classes of functions with bounded, nonintegrable derivatives (see 4.3). Dini did not, at least in print, regard his discovery as indicative of a defect in Riemann's definition, probably because the possibility of an alternative definition had never occurred to him.

In summary, the 1870's mark the period during which Riemann's theory of integration became widely known and accepted. It was primarily a time of accomplishment for the new theory: Du Bois-Reymond proved Fourier's assumption for integrable trigonometric series, despite the fact that term-by-term integration was not considered permissible; Dini's work established the Fundamental Theorem within the more general context of integrable functions; and Riemann's definition of the integral contributed significantly to the overthrow of the belief that continuous functions possess a derivative at all but a few exceptional points. Except for Dini no one attempted to establish anything positive about the differentiability properties of continuous functions, probably because it did not seem that much could be said. And characteristic of mathematicians at this time was their limited understanding of the possibilities of first species sets, nowhere dense sets, and sets of zero content. Indeed, had Dini understood the nature of nowhere dense sets, he probably would have been able to construct an example of a function with nonintegrable derivatives. Dini's conjectures motivated (and were eventually confirmed by) Volterra's later discovery of nowhere dense sets with positive outer content.

3

Set Theory and
the Theory of Integration

During the 1880's research in the theory of integration focused on the properties of infinite sets. The discovery that nowhere dense sets can have positive outer content revealed the importance of measure-theoretic properties of sets in the theory of integration and quickly led to the introduction of the first theory of measure. Set-theoretic considerations also played an important role in attempts to extend the Riemann integral to unbounded functions and revealed the difficulties that beset such extensions—difficulties that focused attention upon the property of absolute continuity and upon the apparent incompatibility of the general theory of curve rectification and the traditional integral formula for curve length. An understanding of these difficulties is particularly important for an appreciation of Lebesgue's contribution to integration theory.

3.1   Nowhere Dense Sets and Their Measure-theoretic Properties

In the *Fondamenti* Dini had rejected Hankel's proof that a pointwise discontinuous function is necessarily integrable, but was unable to construct an example showing that not only the proof but also the proposition itself is erroneous. (Such a counterexample amounts to constructing a nowhere dense set of positive outer content.) Also, in connection with the existence of bounded, nonintegrable derivatives, Dini had posited the existence of continuous, nonconstant functions with densely distributed intervals of invariability and, thereby, the existence of nowhere dense sets of positive outer content, although he was again unable to produce an example.

Volterra's discovery of nowhere dense sets of positive outer content confirmed Dini's opinions. In a paper [1881a] written while he was still a student at the Scuola Normale Superiore di Pisa, Vito Volterra (1860–1940) showed that a nowhere dense set can be constructed by means of the following procedure, which agrees in its essentials with that of H. J. S. Smith (see 2.2): The interval [0,1] is divided into an infinite number of intervals by the sequence $1 > c_1 > c_2 > \cdots$, where: (a) $c_n \to 0$; (b) $1 - c_1 = (1/2^{2 \cdot 1})(1 - 0)$; (c) $c_n - c_{n+1} < 1 - c_1$. The interval $(c_1, 1)$ is excluded from further subdivision, and the same procedure is applied to each of the (c_{n+1}, c_n). That is, let $c_n > c_{n,1} > c_{n,2} > \cdots > c_{n,k} > \cdots$, where: (a) $c_{n,k} \to c_{n+1}$; (b) $c_n - c_{n,1} = (1/2^{2 \cdot 2})(c_n - c_{n+1})$; (c) $c_{n,k} - c_{n,k+1} < c_n - c_{n,1}$. In this manner each of the intervals (c_{n+1}, c_n) is divided into an infinity of intervals $(c_{n,k}, c_{n,k+1})$, for $k = 1, 2, 3, \cdots$. The same procedure is then applied to each of these intervals—the interval $(c_{n,1}, c_n)$ being excluded from further subdivision—and so on *ad infinitum*. Let G denote the set of points of subdivision, that is, the c's together with 0 and 1, and let $\overline{G}$ denote the set of points of G together with all its limit points. G and $\overline{G}$ are then examples of nowhere dense sets.

Volterra showed that $\overline{G}$ has the additional property that if x is a point in [0,1] that is not in $\overline{G}$, then x belongs to an open interval (a,b) whose endpoints a and b belong to $\overline{G}$. (The interval (a,b) is the maximal interval which contains x and is disjoint from $\overline{G}$.) Thus $\overline{G}$ is the complement of a union of disjoint open intervals with endpoints in $\overline{G}$. In fact, these intervals are precisely the intervals which were excluded from further subdivisions in the construction of G, viz., $(c_1,1)$, $(c_{1,1},c_1)$, $\cdots$, $(c_{n,1},c_n)$, $\cdots$. Furthermore, if s_n denotes the total length of those intervals forming $[0,1] - \overline{G}$ that have length greater than or equal to $1/2^{2n}$, then

$$s_n < \frac{1}{2^2} + \frac{1}{2^4} + \cdots + \frac{1}{2^{2n}} < \frac{1}{3},$$

by virtue of (b) and (c). Therefore, if P is any partition of [0,1] of norm less than $1/2^{2n}$, then the total length of those subintervals devoid of points of $\overline{G}$ must be less than s_n and, consequently, less than $1/3$. In other words, $c_e(\overline{G}) > 2/3$. Volterra then considered the function that takes the value 0 on $\overline{G}$ and the value 1 on $[0,1] - \overline{G}$; this function is pointwise discontinuous but fails to satisfy Riemann's integrability condition.

In a later paper Volterra [1881b] made use of the set $\overline{G}$ to confirm another conjecture of Dini's, namely that there exist functions with bounded, nonintegrable derivatives. The idea is to associate with each interval $[a,b]$ a function $f(x;a,b)$ which is defined on $[a,b]$ as follows: (1) $f(x;a,b) = 0$ if $x = a$ or $x = b$; (2) $f(x;a,b) = (x-a)^2 \sin[1/(x-a)]$ when $a < x \leq x_1$, where x_1 is the largest number less than or equal to $(a+b)/2$ for which $(x-a)^2 \sin[1/(x-a)]$ has a zero derivative; (3) $f(x;a,b) = (b-x)^2 \sin[1/(b-x)]$ when $x_2 \leq x < b$, where x_2 is the smallest number greater than or equal to $(a+b)/2$ for which the derivative of $(b-x)^2 \sin[1/(b-x)]$ is zero; (4) $f(x;a,b) = (x_1-a)^2 \sin[1/(x_1-a)] = (b-x_2)^2 \sin[1/(b-x_2)]$ when $x_1 \leq x \leq x_2$. Clearly, $f'(x;a,b)$ behaves in the neighborhood of a or b in the same manner as $[x^2 \sin(1/x)]' = 2x \sin(1/x) - \cos(1/x)$ behaves in the neighborhood of 0. That is to say, as x approaches either a or b, $f'(x;a,b)$ oscillates indefinitely between -1 and 1. Furthermore, $f(x;a,b)$ has the following properties:

$$\left| f(x;a,b) \right| < \begin{cases} (b-x)^2 \\ (x-a)^2 \end{cases}$$

$$\left| f'(x;a,b) \right| < 2(b-a) + 1.$$

Making use of these two properties, Volterra showed that if a function g is defined on $[0,1]$ by setting $g(x)$ equal to $f(x;a,b)$ on each of the intervals (a,b) comprising the complement of $\overline{G}$ and equal to 0 on $\overline{G}$, then g is everywhere differentiable and $\left| g'(x) \right| \leq 3$ for all x in $[0,1]$. Since, however, the function g' has an oscillation equal to 2 at the points of G—the endpoints of the intervals forming the complement of $\overline{G}$—g' is not Riemann-integrable.

The function g thus shows that "in some cases, it can happen that the ordinary definition of the integral [in terms of a primitive function] is not included in that of Riemann ... " [Volterra 1881b: 334]. But this remark was not seriously intended as a criticism of Riemann's definition; Volterra hastened to add that Riemann's definition, because it established the existence of the integral, was superior to the older conception, which posited the existence of a primitive of the function to be integrated.

Volterra also pointed out that if f denotes the characteristic function of the set $[0,1] - \overline{G}$—i.e., the example of a nonintegrable function considered earlier—then the functions $F(x) = \overline{\int_0^x} f$ and $G(x) = \underline{\int_0^x} f$ are examples of Dini's continuous functions with densely distributed intervals of invariability. Thus, the Dini derivatives of F and G (D^+F, D^+G, etc.) are bounded but not integrable. For functions such as g, F, and G, the Fundamental Theorem II consequently loses its meaning, as Dini had foreseen.

The discovery of nowhere dense sets of positive outer content brought about the recognition of the importance for the theory of integration of the measure-theoretic properties of sets. This recognition is evidenced to some extent in Volterra's papers, but in Germany, where the same discovery was made independently of Volterra and Smith, it was emphasized to such a degree that a theory of measure was created. The first German to call attention to nowhere dense sets of positive outer content was Paul du Bois-Reymond [1880: 127]. By confounding nowhere dense and first species sets, du Bois-Reymond had come to regard Dirichlet's integrability condition—that the points of discontinuity of a function form a nowhere dense set—as a special case of Riemann's. Later, however, he discovered that nowhere dense sets could possess what he called condensation points of infinite order. Such a set can be constructed in the following manner:[1] Let p denote a fixed real number, and let I_1, I_2, $\cdots$, I_n, $\cdots$ denote a sequence of disjoint intervals with endpoints converging to p. Define a first species set P_n of type n in the interval I_n (P_n is of type n if its nth derived set, $P_n^{(n)}$, is finite). Then if $P = \bigcup_{n=1}^{\infty} P_n$, the point p is a condensation point of infinite order with respect to P in the sense that p belongs to $P^{(n)}$ for $n = 1, 2, 3, \cdots$. Therefore P is not a first species set, although it is clearly nowhere dense.

The idea behind the construction of P (in contrast to the construction of first species sets) is to distribute intervals rather than just points. That then suggested to du Bois-Reymond that a nowhere dense set could be constructed by densely distributing an infinite number of disjoint intervals in a fixed interval such as $[-\pi, \pi]$; the points of $[-\pi, \pi]$ which are not in these intervals will form a nowhere dense set Q. Furthermore, if the intervals are determined in such a manner that the sum of their lengths is less than 2π, Q can not be enclosed in a finite number of intervals of arbitrarily small total length. Thus, as du Bois-Reymond pointed out, the characteristic function of Q is not Riemann-integrable, although it satisfies Dirichlet's condition.

The manner of actually effecting the dense distribution of infinitely many disjoint intervals was made more explicit in du Bois-Reymond's *Die allgemeine Functionentheorie* [1882] and agrees in essentials with the methods of Smith and Volterra, whose papers were still not known in Germany. Similar constructive definitions were also published in 1882 by W. Veltmann [1882] and Axel Harnack [1882a: 239]. Du Bois-Reymond [1882: 188–89] introduced the term "integrable system of

---

1. du Bois-Reymond 1880: 127n; cf. 1882: 187–88.

points" to refer to sets of content zero to distinguish them from other nowhere dense sets. But it was Axel Harnack who developed the measure-theoretic notion of sets of content zero and gave it prominence in the periodical literature of the early 1880's.

In his paper on trigonometric series [1880], Harnack had failed to distinguish sets of zero content and first species sets (see 2.2). By the time he published *Die Elemente der Differential- und Integralrechnung* [1881], this distinction had been made, and the first species sets of his 1880 definition—sets of zero content—were now termed "discrete sets," since, for the problems of the integral calculus, they are of the same nature as a finite number of separated points [1881: 259]. Cantor [1880: 358] had constructed a set similar to du Bois-Reymond's set P with a condensation point of infinite order. This example of a discrete set not of the first species may have indicated the distinction to Harnack, for he gave it in *Die Elemente*. Thus, fourteen years after Riemann's paper appeared in print, the measure-theoretic element implicit in integrability condition (R_2) finally became conceptualized. Hankel's theorem in rectified form now reads: $f(x)$ is integrable if and only if for every positive σ, S_σ, the set of x at which the oscillation of the function is greater than σ, is discrete.

In a subsequent paper Harnack [1882a] introduced a notion analogous to that of a property holding "almost everywhere." Two functions f and g are said to be equal "in general" if, for every $\delta > 0$, the set of points x such that $|f(x) - g(x)| > \delta$ is discrete. Thus, for example, Dini's version of the Fundamental Theorem I becomes: If $F(x) = \int_a^x f$ and DF denotes one of the Dini derivatives of F, then $DF(x) = f(x)$ "in general." The notion of "in general"—like Dini's functions with null integrals[2]—thus plays the same role within Riemann's theory of the integral as "almost everywhere" does in Lebesgue's. In fact, it is clear that the Fundamental Theorem I could have been expressed as: $DF(x) = f(x)$ almost everywhere. Likewise, Riemann's integrability condition is equivalent to the condition that a bounded function be continuous almost everywhere. The eventual discovery of this equivalence did in fact help convince Giuseppe Vitali and W. H. Young— independently of each other and of Lebesgue—of the value of a theory of measure (in the sense of Lebesgue). But in Harnack's case there were additional reasons for sticking with measure-theoretic ideas based on content (see 3.2).

2. A proof that f is a function with null integrals if and only if $f = 0$ "in general" was given by M. Pasch [1887].

By developing the consequences of the concept of a discrete set, Harnack believed he would be in a position to state all the theorems of the integral calculus succinctly and with the greatest possible degree of generality. For example, Dini had proved in his *Fondamenti* that two integrable functions have the same integral if they agree except on a first species set; but Harnack was able to show that two integrable functions have the same integral if and only if they are equal "in general." Harnack [1882a: 241] also thought he had established a result which implies the following generalization of the Fundamental Theorem III: Suppose f is continuous and $f'(x) = 0$ "in general" (in the sense that for every $\epsilon > 0$ there exists a discrete set A_ϵ such that if x is not in A_ϵ, then a $\delta = \delta(x)$ can be determined so that $\left| \dfrac{f(x+h) - f(x)}{h} \right| < \epsilon$ when $|h| < \delta$); then f must be a constant function. The theory of discrete sets in general and this (false) proposition in particular were the means by which he hoped to simplify and extend the theory of trigonometric series. The vast majority of his results eventually proved to be incorrect, but his papers are important because they drew attention to the theory of discrete sets.

Harnack was especially fascinated by the inequality

$$\tfrac{1}{2} a_0{}^2 + \sum_{n=1}^{N} (a_n{}^2 + b_n{}^2) \leq \int_{-\pi}^{\pi} f^2,$$

where $a_0, a_1, \cdots, a_N, b_1, b_2, \cdots, b_N$ denote the Fourier coefficients of the square-integrable function f. This inequality (usually referred to now as Bessel's inequality) shows that the series $\sum_{n=1}^{\infty} (a_n{}^2 + b_n{}^2)$ converges; hence, if

$$S_n(x) = \tfrac{1}{2} a_0 + \sum_{k=1}^{n} (a_k \cos kx + b_k \sin kx),$$

then the integral

$$\int_{-\pi}^{\pi} (S_{n+p} - S_n)^2 = \sum_{k=n}^{n+p} (a_k{}^2 + b_k{}^2)$$

converges to 0 as $n \to \infty$. Harnack believed that he could use this property of the S_n to deduce that $g(x) = \lim_{n \to \infty} S_n(x)$ must exist "in general,"[3]

---

3. That is, for every positive ϵ the set of x such that $\limsup_{n \to \infty} S_n(x) - \liminf_{n \to \infty} S_n(x) \geq \epsilon$ is discrete.

that

$$\int_{-\pi}^{\pi} g^2 = \tfrac{1}{2} a_0^2 + \sum_{n=1}^{\infty} (a_n^2 + b_n^2)$$

and that, in fact, $\int_{-\pi}^{\pi} (f-g)^2 = 0$, so $f(x) = g(x)$ "in general."

What is particularly interesting about these "results" is that they are not far removed from the truth, since if S_n is a sequence of square-integrable functions such that $\int_{-\pi}^{\pi} (S_n - S_m)^2 \to 0$ as $n,m \to \infty$, then there exists a limit function g in the sense that $\int_{-\pi}^{\pi} (g - S_n)^2 \to 0$ as $n \to \infty$; moreover, for a suitably chosen subsequence n_k, $g(x) = \lim_{k \to \infty} S_{n_k}(x)$ almost everywhere. It is crucial, however, that the integrals be taken in the sense of Lebesgue (see Chapter 6).

The same year that Harnack's paper appeared, G. Halphén [1882] also observed that for square-integrable functions, the Fourier coefficients converge to zero, since $\sum_{n=1}^{\infty} (a_n^2 + b_n^2)$ converges. But he also pointed out, by an example involving Tchebycheff polynomials, that $\lim_{n,m \to \infty} \int_a^b [S_n(x) - S_m(x)]^2 dx$ can be 0 while $\lim_{n \to \infty} S_n(x)$ fails to exist for all but one value of x. Unaware of Halphén's counterexample, Harnack [1882b] soon admitted the untenability of his conclusions because his limit function $g(x)$ need not be Riemann-integrable as he had tacitly assumed.

3.2   THE INTRODUCTION OF OUTER CONTENT

The first mathematician to introduce the definition of the outer content of a set was Otto Stolz (1842–1905), Professor of Mathematics at the University of Innsbruck. In 1881, when working on the problem of defining curve length, he expressed the view that area and volume should be defined in terms of the integral. Harnack's work on discrete sets prompted Stolz to place them within the context of a general theory of content.[4] Stolz's contribution [1884] amounts to observing that if E is an arbitrary subset of some interval $[a,b]$, then E can be

---

4.   It should be noted that Stolz was the reviewer in *Fortschr. Math.* of Harnack's papers.

assigned a number L with the following property: for every $\epsilon > 0$ there exists a $\delta > 0$ such that if P is a partition of $[a,b]$ and $\|P\| < \delta$, then $|L(P) - L| < \epsilon$, where $L(P)$ denotes the sum of the lengths of the intervals determined by P which contain points of E. In particular, $L = 0$ if and only if E is discrete. Stolz also extended his definition of L to bounded subsets in the plane.

Later in 1884, and independently of Stolz, Cantor [1884: 473–79] published an equivalent definition of content—he also introduced the term "content" (*Inhalt*)—but within the more general context of n-dimensional Euclidean space. Cantor's definition is based upon two assumptions, to which he failed to pay sufficient attention. First, he assumed that if P is composed of a finite number of bounded, n-dimensional "regions" (*Stücken*), then the multiple integral $\int_P dx_1 dx_2 \cdots dx_n$ is defined. Secondly, he assumed that if P is an arbitrary bounded set, then for any positive r the set $\Pi(P,r) = \bigcup_{p \in \overline{P}} K(p,r)$ consists of a finite number of n-dimensional pieces, where $\overline{P}$ denotes P together with its limit points, and $K(p,r)$ denotes the sphere of radius r and center p. The first assumption guarantees that the integral

$$\int_{\Pi(P,r)} dx_1 dx_2 \cdots dx_n$$

is defined. But, before Jordan's work in 1892 (cf. Ch. 4), the theory of multiple integrals was not treated with the generality or precision assumed for it in this context by Cantor. The second assumption can easily be justified from the fact that the set $\overline{P}$, being closed and bounded, is compact. But this, too, was not discovered for some time to come.

Cantor then defined $I(P)$, the content or volume of P, by

$$I(P) = \lim_{r \to 0} \int_{\Pi(P,r)} dx_1 dx_2 \cdots dx_n.$$

This definition makes it clear that, in the determination of content, no distinction is made between a set and its closure—they are assigned the same numerical measure. Cantor was the first to show some interest in the additive properties of the set function $I(P)$. The definition makes it clear that for two disjoint sets P and Q, the relation $I(P \cup Q) = I(P) + I(Q)$ need not hold if their closures are not likewise disjoint. (Indeed, if $\overline{P} = \overline{Q} = P \cup Q$, then $I(P) + I(Q) = 2I(P \cup Q)$.) If, however, P and Q are contained in completely separated n-dimensional portions of space, then, as Cantor noted, $I(P \cup Q) = I(P) + I(Q)$.

It appears fairly evident that the primary motivation behind Cantor's definition came from du Bois-Reymond's and Harnack's discussions of sets of null content, a topic Cantor [1883: 54] had earlier shown an interest in. But he also had in mind the possibility of the future application of the notion of content to the study of continua. When he had presented a precise definition of a continuum in an earlier paper [1883: 590n12], he had called attention to the independence of the definition from dimension considerations and promised a subsequent study to show that continua could be further classified by dimension. Cantor expanded on this remark in a French version of the paper containing his definition of content. The content of a set P, he stressed, is completely dependent on the dimension of the space E^n in which it is considered to be a subset. That is, if P is also considered as a subset of E^m, the m-dimensional content of P will generally differ from its n-dimensional content; a unit square considered as a subset of 3-dimensional space has content 0, and considered as a subset of 2-dimensional space has content 1. "This general notion of *volume* or *magnitude* is indispensable to me in investigations on the *dimensions* of *continuous sets* . . . " [1884b: 389–90]. Unfortunately, Cantor never developed these ideas further.[5] But the applications envisioned by him for the theory of content were of such a nature that an outer content or measure was appropriate. It was primarily in connection with the theory of multiple integrals over general sets that measurability became relevant, and Cantor had glossed over this subject.

Without the theorem that a closed and bounded subset of n-dimensional space is compact, Cantor's definition does not appear at first to be a faithful generalization of the notion of a discrete set, as the set to be measured, $\overline{P}$, is covered by an infinite number of spheres. In a paper submitted shortly after the appearance of Cantor's, Harnack [1885] proposed his own definition of the content of a set. In the definition of a discrete set, Harnack pointed out, the set is covered by a finite number of intervals; consequently he proceeded to define the content of a bounded subset of the real numbers as the limit of the sum of the lengths of finite interval covers. (Harnack was apparently unaware of Stolz's similar definition.)

Having given his own definition of content, Harnack proceeded to show that the theorems proved by Cantor retain their validity with his

5. The reason for this is probably that Cantor soon became completely absorbed with the theory of transfinite numbers. A metric approach to dimension was also developed later by F. Hausdorff [1919], who based it upon Carathéodory's definition of the m-dimensional outer measure of a subset of an n-dimensional space [1914].

definition. Cantor's definition also turned Harnack's interest to the question of what happens when an infinite number of covering intervals is allowed in the definition of content. On the one hand, he observed, a countable union of discrete sets need not be discrete, as the set of rational numbers in [0,1] illustrates. But "in a certain sense every 'countable' point set has the property that all its points can be enclosed in intervals whose sum is arbitrarily small" [1885: 243]. That is, he continues, if $a_1, a_2, \cdots, a_n, \cdots$ denote the members of a countable set and if $\delta > 0$ is given, each a_n can be enclosed in an interval of length ϵ_n, where $\delta = \sum_{n=1}^{\infty} \epsilon_n$. Harnack appears to have been the first mathematician to call attention to this property of countable sets, the property which was later to suggest to Borel the usefulness of a theory of measure. It may have been suggested to Harnack by Hankel's discussion of the special countable set of points $1/2^n$, in which the same technique is employed (see 2.2).

The point Harnack made, of course, is that since countable sets can be dense and therefore need not be discrete, the restriction to a finite number of covering intervals in the definition of content is crucial. The earlier belief that topologically negligible sets (nowhere dense sets) should be negligible in a measure-theoretic sense had been abandoned, but the feeling persisted that topologically large sets (everywhere dense sets) should not have negligible measure. The fact that certain everywhere dense sets can be enclosed in intervals of arbitrarily small total length—a fact now so familiar as to appear commonplace—must have seemed too paradoxical to serve as the basis for a theory of measure.

Harnack had another reason for doubting the fruitfulness of definitions of content employing infinite interval covers: He thought he had discovered another paradox that arises when an infinite number of intervals are involved—that it is possible to express an interval $[a,b]$ as a union of an infinite number of nonoverlapping intervals in such a way that the sum of their lengths is less than the length of $[a,b]$. Harnack was led to this "paradox" through his investigation of the relationship between his definition of content and the use of an infinite number of covering intervals. His investigation took the form of the following question: If a set of points is enclosed in an infinite number of nonoverlapping intervals I_i with length ϵ_i, and if all these intervals are contained in a fixed interval $[a,b]$ of length L so that $s = \sum_{i=1}^{\infty} \epsilon_i$ is less than L, "under what conditions can it be concluded that they can also be enclosed in a *finite* number of intervals whose sum differs arbitrarily little from s?" [1885: 243]. At first glance it might appear that Harnack was asking for conditions under which the underlying set is compact, but the question, as he interpreted it, is actually slightly different: Given a set E which is the sum of nonoverlapping intervals I_i, when is it true that

$$(7) \qquad c_e(E) = \sum_{i=1}^{\infty} \epsilon_i?$$

This question is particularly interesting and important because Borel later took the right-hand side of (7) as the definition of the measure of E, regardless of whether it coincides with $c_e(E)$.

It should be noted that the boundary of E, ∂E, is the closure of the endpoints of the I_i.[6] Thus, since

$$c_e(E) = c_i(E) + c_e(\partial E) = \sum_{i=1}^{\infty} \epsilon_i + c_e(\partial E),$$

(7) is true if and only if the set of endpoints of the I_i has outer content zero. Harnack was unable to see that his problem could be given such a simple answer. Instead, he imposed the rather strong and unnecessary condition on E that:

(8) $\quad [a,b] - E$ *is also expressible as a union of nonoverlapping intervals* J_i.

Assuming (8), it then follows easily that (7) is true if and only if

$$(9) \qquad \sum_{i=1}^{\infty} \eta_i = L - s,$$

where η_i denotes the length of J_i. Next Harnack showed that (9) is equivalent to:

(10) *The limit points of the endpoints of the J_i must form a discrete set.*

To prove that (9) implies (10), he observed that if s_1 denotes the (outer) content of these endpoints, then

$$\sum_{i=1}^{\infty} \eta_i = L - s - s_1.$$

---

6. The boundary of E consists of the boundary points of E. The latter are those points x such that every open interval containing x contains points belonging to E as well as points not in E. This notion was introduced by Peano and Jordan. (See 4.1.)

(This follows from the fact that $L-s=c_e\left(\bigcup_{i=1}^{\infty}J_i\right)=c_i\left(\bigcup_{i=1}^{\infty}J_i\right)+s_1=$

$\sum_{i=1}^{\infty}\eta_i+s_1$; Harnack did not reveal his own reasoning.)

Harnack then returned to the case in which (10) is not satisfied. In this case—with condition (8) still assumed—$[a,b]$ is decomposed into nonoverlapping intervals $I_1, I_2, I_3, \cdots, J_1, J_2, J_3, \cdots$ in such a manner that

$$\sum_{i=1}^{\infty}\epsilon_i + \sum_{i=1}^{\infty}\eta_i = s + L - s - s_1 = L - s_1 < L.$$

Thus Harnack thought he had discovered another paradoxical consequence of the passage from the finite to the infinite. Such a decomposition as he imagined is, of course, impossible, because Lebesgue measure is countably additive. Not having the theory of measure at his disposal, Harnack probably regarded his discovery as further reason to reject the idea that the content of an infinite union of nonoverlapping intervals should always equal the sum of their lengths.

The content introduced independently by Stolz, Cantor, and Harnack is an outer content, as it must be to relate to the theory of discrete sets and thereby to Riemann's integrability condition (R_2). Consequently, the definition is applicable to every bounded subset of the plane. In particular, if f is defined and bounded on $[a,b]$ and if E denotes the set of points in the plane bounded by the graph of f, the lines $x=a$ and $x=b$, and the x-axis, then E and the subsets E^+ and E^- (those points of E above and below the x-axis) possess contents whether or not f is integrable. Thus, the customary relationship between the concepts of area and the integral, i.e.,

$$\int_a^b f = \text{area } (E^+) - \text{area } (E^-), \qquad \int_a^b |f| = \text{area } (E),$$

is no longer valid if "area" is identified with "content."

In his concluding remarks, Harnack turned to this point and observed that if f is a positive function (so that $E = E^+$), then the content of E cannot be expressed by an integral unless the boundary of E, which he left undefined, is a discrete set—unless, in other words, E is measurable in Jordan's sense. But it never occurred to him to restrict content to this type of set just as integration is restricted to integrable functions. After all, what reason was there for wishing to restrict the generality of the theory of content? It was primarily through the theory of multiple integrals and, in particular, through the work of Camille Jordan that the importance of the notion of measurability was first recognized.

The discovery of nowhere dense sets with positive outer content had revealed the existence of an unlimited number of bounded functions not integrable in Riemann's sense and an equally vast number of functions with bounded, nonintegrable derivatives. These results, however, did not suggest to mathematicians that Riemann's definition was too restrictive. On the contrary, it was accepted, as Weierstrass observed, as "the most general conceivable" [Mittag-Leffler 1923: 196]. Prior to Lebesgue, the single critic of Riemann's definition was Karl Weierstrass. His dissatisfaction and the alternate theory of integration he proposed are particularly relevant, because Weierstrass attempted to extend the definition of the integral by means of Cantor's theory of content.

Judging by some of the comments made by Weierstrass in his correspondence, he had never been satisfied with Riemann's definition of the integral. In a letter of April 1885 he shared his dissatisfaction with Paul du Bois-Reymond, who had written of his discovery that Dirichlet's condition for integrability is not sufficient for Riemann integrability. Weierstrass responded by astutely observing that when Dirichlet formulated his condition, he undoubtedly had in mind a different conception of the definite integral than that later proposed by Riemann. What Dirichlet had actually intended, Weierstrass suggested, was to extend Cauchy's definition to functions with an infinite number of points of discontinuity by considering the integral as the sum of the integrals taken over all maximal intervals of continuity of the function. Therefore, to say that Dirichlet's condition is not sufficient for integrability, one must first accept Riemann's conception of integrability. And Weierstrass had his reasons for finding it unacceptable:

> It has always seemed to me that Riemann's definition of the definite integral is subject to a disadvantage which I had to ignore since till now I did not know how to remove it. It is this: in determining the greatest oscillation that the value of the function to be integrated makes in an interval (therefore in the statement of the integrability condition), those values it has at the points of discontinuity are considered while, however, the value of the integral, when it exists, depends solely on the values which the function takes at the points of continuity. [1923a: 215.]

In other words, since, as Hankel proved, the points of continuity of an integrable function must form a dense set, then corresponding to any partition of $[a,b]$, the Cauchy sum $S = \sum f(t_i)(x_i - x_{i-1})$ can always be chosen so that the t_i are continuity points of f.

Weierstrass proposed removing this "incongruity" by replacing Riemann's condition (R_1) with the requirement that

$$\lim_{\|P\| \to 0} (D_1{}^*\delta_1 + D_2{}^*\delta_2 + \cdots + D_n{}^*\delta_n) = 0,$$

where $D_i{}^*$ denotes the oscillation of f on the points of continuity of the ith subinterval determined by the partition P. This is equivalent to the condition that $\underline{\int}_C f = \overline{\int}_C f$, where C denotes the set of points of continuity of f; presumably the integral of f was to be defined as this common value.[7] Weierstrass regarded this definition as a generalization of what Dirichlet had in mind. But he failed to realize that, unlike Dirichlet's condition, his still imposed a restriction upon the nature of the discontinuities of the function. Volterra's example of a nonintegrable derivative $g'(x)$, for example, remains nonintegrable in Weierstrass' sense because the intervals of a partition P on which the oscillation of g' is 2 still yield an oscillation of 2 when the points of discontinuity are removed from consideration. Nevertheless, g' satisfies Dirichlet's condition.

In his next letter du Bois-Reymond pointed out this oversight with a counterexample which may have been Volterra's function g'. Weierstrass was then convinced that "the Riemannian definition must be modified more drastically than I have done, i.e. further freed from inessentials . . ." [1923a: 218]. In fact, du Bois-Reymond's counterexample, Weierstrass pointed out, actually supported that contention because it could be represented by a Fourier series if the coefficients a_n and b_n are interpreted as integrals in Dirichlet's sense. (Lebesgue [1903a: 481 ff] proved that Volterra's g' can be represented by a Fourier series in the context of Lebesgue integrals.) It seemed desirable to have a more general definition of the integral—one that would include Dirichlet's as well as Riemann's and thus extend the validity of Fourier's assumption that the coefficients in a trigonometric series representation are integrals.

7. For an arbitrary set F, $\overline{\int}_F f$ is defined as $\overline{\int}_a^b g$, where $g(x) = f(x)$ for x in F and $g(x) = 0$ otherwise.

At this time Weierstrass had already formulated another definition, which made the class of integrable functions coextensive with the class of pointwise discontinuous functions, but he did not disclose it in his letter. Within a few weeks he indicated to his friend and former student, Sonja Kowalewsky, that he had succeeded in extending a definition of the integral to any function that is defined and bounded on a dense subset of an interval. In this form it was included in his lectures on the theory of functions during the summer of 1886:

> Let an interval $(a \cdots b)$ be given, and in every arbitrarily small part of $(a \cdots b)$ let points exist at which the function is defined. Then at every point where the function is defined, I erect the ordinate. These ordinates certainly lie infinitely close to one another, although they need not succeed each other continuously. Therefore $\int_a^b f(x)dx$ cannot be defined as the area filled up by the ordinates, as is possible and permissible for continuous functions. We now make the following definition: Imagine each of the existing ordinates surrounded by a small rectangle whose base is equal to δ. Then these rectangles overlap. If we now define the point set consisting of those points that lie within some such rectangle, it is easily seen that this forms a continuum. This continuum has a content S_δ which is a function of δ .... It can now be shown that S_δ decreases with decreasing δ and, consequently, approaches a limit for $\delta = 0$. We now define $\int_a^b f(x)dx = \lim_{\delta=0} S_\delta$. This definition is justified since it agrees with the familiar one for continuous functions, and also the essential properties of the integral continue to hold. [1923a: 225, *n*16.]

It is not difficult to see that if a function f is defined on a dense subset, F, of $[a,b]$, then (with W for "Weierstrass")

$$(11) \qquad (W) \int_a^b f = \int_F \overline{f}.$$

Indeed (11) indicates that Weierstrass could have defined $(W) \int_a^b f$ directly in terms of Cantor's content. If, for example, $f(x)$ is nonnegative for all x in F, then $(W) \displaystyle\int_a^b f = c_e(E)$, where E is the set of points (x,y) such that x belongs to F and $0 \leq y \leq f(x)$. Weierstrass undoubtedly realized this but did not wish to assume a knowledge of Cantor's theory of content on the part of his audience. The introduction of S_δ is clearly analogous to Cantor's introduction of the content of the set $\Pi(P,r)$.

Weierstrass' definition of $\displaystyle\int_a^b f(x)dx$ as $\lim_{\delta=0} S_\delta$ probably indicates what he had in mind earlier when he wrote to du Bois-Reymond that he had formulated a definition applicable to all bounded, pointwise discontinuous functions. For if f denotes such a function, its points of continuity form a dense set C; the behavior of f at its points of discontinuity, Weierstrass realized, can be entirely disregarded by his definition when F is taken as the set C. Then, because his definition remains meaningful for any function defined and bounded on a dense set F, Weierstrass would have arrived at his final definition.

Weierstrass emphasized that besides coinciding with the Cauchy definition of the integral for continuous functions, a generalized definition should preserve the "essential properties." But what did he understand by "essential"? This remains unclear because there is no evidence of an exhaustive list of these properties. The question of essential properties, a question that becomes of crucial importance with Borel and Lebesgue, was foreign to the thinking of analysts at this time. Nevertheless, certain basic properties were usually singled out in expository treatments of the Riemann integral; among these, the additivity property

$$\int_a^b (f+g) = \int_a^b f + \int_a^b g$$

was invariably included. Because Weierstrass' integral is actually an upper integral, it fails to have this property (as Volterra had pointed out in 1881). It remains unclear whether Weierstrass was cognizant of this fundamental drawback of his definition.

The approach taken by Weierstrass—to extend the definition of the integral by considering it as the area of the ordinate set—was a step in the right direction, but the theory of area that he adopted was not sufficiently refined to impart the essential property of additivity to the resulting integral. Historically it was the introduction of the concept of measurability in the work of Peano and Jordan that was to suggest the manner of insuring the requisite refinement.

3.3 Cantor's Development of the Theory of Sets and its Application to the Theory of Integration

Cantor's introduction of the notion of the content of a set occurred within the context of a much more far-reaching study of the properties of infinite sets of points, published in installments during the years 1879–84 under the title "Ueber unendliche, lineare Punktmannichfaltigkeiten." Infinite sets had already played an important role in Cantor's 1872 work on trigonometric series (2.1). The following year he introduced the concept of a countable set and proved the striking theorem that the set of all algebraic numbers is countable [1874]. Then in "lineare Punktmannichfaltigkeiten" Cantor set out to classify and study infinite sets of points by means of the notions of countability and uncountability and the properties of the derived sets of a given set. His results are significant in the historical development of the theory of integration.

In the first installment Cantor [1879] introduced the term "first species set" for those sets he had originally referred to as of the nth type. Here also the terms "everywhere dense set" and "nowhere dense set" were used for the first time, although the properties behind these terms had been under discussion for some time. In fact, Cantor was undoubtedly encouraged to investigate further the properties of sets because of their relevance to the theory of functions of a real variable, particularly to the theory of integration; in this connection he refers to Dini's use of first species sets in the *Fondamenti*.

In the second installment Cantor [1880] introduced the notions of set union and intersection and used them to extend the concept of the nth derived set, $P^{(n)}$, of a set P: If P is not of the first species, then $P^{(\infty)}$ is defined to be $\bigcap_{n=1}^{\infty} P^{(n)}$. Once $P^{(\infty)}$ has been defined, $P^{(\infty+n)}$ can be defined naturally as $[P^{(\infty)}]^{(n)}$ for $n = 1, 2, 3, \cdots$. Then $P^{(2\infty)}$ is defined as $[P^{(\infty)}]^{(\infty)}$, $P^{(3\infty)}$ as $[P^{(2\infty)}]^{(\infty)}$, and so on. Next, $P^{(\infty^2)}$ is defined to be $\bigcap_{n=1}^{\infty} P^{(n\infty)}$ and $P^{(\infty^\infty)}$ as $\bigcap_{n=1}^{\infty} P^{(\infty^n)}$. Furthermore, this "dialectical generation of concepts" can be continued indefinitely so that, in particular, $P^{(\gamma)}$ can be defined for every "infinity symbol" γ of the form

$$\gamma = n_k \infty^k + n_{k-1} \infty^{k-1} + \cdots + n_1 \infty + n_0,$$

where n_0, n_1, $\cdots$, n_k denote integers. Here we have the beginnings of Cantor's theory of transfinite numbers, to which he was later to devote almost all his attention. (See [Jourdain 1910; 1913a; 1913b].)

Consideration of the sets $P^{(\gamma)}$ enabled Cantor in a subsequent install-ment [1883b] to express the derived set of an arbitrary set P as $P' = R \cup S$, where $R = \bigcup_\gamma [P^{(\gamma)} - P^{(\gamma+1)}]$ and $S = \bigcap_\gamma P^{(\gamma)}$. The union and intersection are taken over all γ belonging to what Cantor called the first two *Zahlenclassen*, that is, over all ordinal numbers γ less than Ω, where Ω is the first ordinal corresponding to an uncountable well-ordered set.[8] The set S has the property that $S' = S$, and Cantor intro-duced the term "perfect set" for sets with this property.[9] Cantor proved that R is countable and also claimed a bit too hastily that R is also a "reducible set" —i.e., that $R^{(\gamma)} = \emptyset$ for some γ less than Ω. In a letter to Cantor, Ivar Bendixson (1861–1936), a young Swedish mathemati-cian, showed that although R will always be countable, it need not be reducible [1883]. Cantor [1884a: 470] was thus led to introduce the con-cept of a closed set, a set P being defined to be closed if and only if P' is a subset of P. The relevance of closed sets to Cantor's problem is re-vealed by the proposition, which he proved, that a set R is closed if and only if $R = Q'$ for some set Q. Cantor had already proved that any set Q with a countable derived set Q' is reducible. Thus, any closed, count-able set must be reducible, and, in particular, the set R in the decompo-sition $P' = R \cup S$ will be reducible provided R is closed. (An immediate corollary to this result is the proposition that a nonempty perfect set is uncountable.)

The introduction of the concept of a closed set becomes important in connection with Borel's theory of measure because of another result of Cantor's. In an earlier part of "lineare Punktmannichfaltigkeiten," Cantor [1883a] was prompted by the work of du Bois-Reymond and Harnack to consider the problem of determining conditions under which a nowhere dense set has zero content. At that time he was able to prove that if P' is countable, then P has zero content. It is his method of proof that is of interest: Let $Q = P \cup P'$ and assume (without loss of generality) that Q is a subset of $[0,1]$ and that 0 and 1 are in Q. Then, if

---

8. Cantor actually did not explain how R and S are defined until the next installment, but it is fairly clear that this is what he had in mind originally. Of course, he had not yet developed the notion of an ordinal number.

9. It is in this paper that Cantor [1883b: 590] introduced the now-familiar Cantor ternary set; but, as we have seen, "Cantor sets" were not discovered by Cantor.

r is a point in $[0, 1]-Q$, r cannot be a limit point of Q; hence r is contained in an interval (c,d) which does not contain any points of Q. If (c,d) is maximal with respect to these properties, then c and d will belong to Q. It thus follows that $[0,1]-Q$ can be represented as $\bigcup_{n=1}^{\infty} (c_n, d_n)$, a union of nonoverlapping intervals with endpoints in Q. Cantor then went on to show that the sum of the lengths of these intervals is 1, so Q and therefore P must have a zero content. What is important here is the characterization of Q as the complement of nonoverlapping intervals. (This is similar to Volterra's characterization of his set $\overline{G}$ (3.1), although Cantor probably did not read Volterra's paper.) In his letter to Cantor, Bendixson made use of the fact that any perfect nowhere dense S can be characterized as the complement of nonoverlapping intervals and that, in fact, S is the derived set of the set of endpoints c_n and d_n. Once the notion of a closed set was introduced, it was clear that any closed set can also be considered as the complement of nonoverlapping intervals, thus suggesting that the measure of a closed set Q in $[0,1]$ should be $1 - \sum_{n=1}^{\infty} (d_n - c_n)$. In fact it turns out that $c_e(Q) = 1 - \sum_{n=1}^{\infty} (d_n - c_n)$, so this is in agreement with the definition of measure accepted at the time. The assignment of this measure to Q, however, also suggests defining the measure of the complement of Q, $R = \bigcup_{n=1}^{\infty} (c_n, d_n)$, as $\sum_{n=1}^{\infty} (d_n - c_n)$ so that the measures of Q and R add up to 1. This is what Borel proposed (4.2).

One of Cantor's major objectives was to show that every set of points is either countable or has the power of the continuum $[0,1]$. As a first step towards a complete solution, he proposed to show that every perfect set can be put in one-to-one correspondence with $[0,1]$. The problem clearly reduces to establishing that every nowhere dense perfect set can be put in one-to-one correspondence with $[0,1]$—a fact Cantor succeeded in proving by making use of the above-mentioned characterization of these sets. Specifically, Cantor assumed without loss of generality that S is a perfect nowhere dense subset of $[0,1]$, so $[0,1]-S = \bigcup_{n=1}^{\infty} (a_n, b_n)$, and that 0 and 1 are among the a_n and b_n. Thus S can be expressed in the form $\{a_n\} \cup \{b_n\} \cup G$. Cantor's idea was to show that if D denotes any countable dense subset of $[0,1]$, then the points of D can be ordered as a sequence $d_1, d_2, d_3, \cdots$ in such a manner that the correspondence $d_n \leftrightarrow (a_n, b_n)$ is order-preserving. That is, if

(a_n, b_n) is to the left of (a_m, b_m), then d_n is to the left of d_m. Once this was established, it remained to show that the two uncountable sets G and $F = [0,1] - D$ can also be put in one-to-one correspondence with each other. Now if f is a member of F, then, since D is dense, there exists a subsequence d_{n_i} of D which converges to f. Because the correspondence $d_n \leftrightarrow (a_n, b_n)$ preserves order, the sequence (a_{n_i}, b_{n_i}) must converge to a point g in G. It then follows that S has the power of the continuum because the correspondence $f \leftrightarrow g$ is one-to-one between F and all of G.

In a revised, French version of this proof, Cantor added the observation that the proof has the advantage of revealing an extensive and remarkable class of continuous functions whose properties seemed to suggest interesting avenues for further research [1884b: 385]. These functions arise from the perfect set S as follows: Define φ on $[0,1]$ by $\varphi(g) = f$ for each g in G, where $g \leftrightarrow f$ by the one-to-one correspondence between F and G; φ can be easily extended to all of $[0,1]$ because

$$[0,1] - G = \bigcup_{n=1}^{\infty} [a_n, b_n],$$ so for all x in $[a_n, b_n]$, $\varphi(x)$ can be defined as d_n,

where $d_n \leftrightarrow (a_n, b_n)$ in the order-preserving correspondence. Then φ is continuous and monotonically increasing. Furthermore, since φ is constant on the intervals (a_n, b_n), it follows that $\varphi'(x)$ exists for all x not in S and has the value 0. When S is discrete— i.e., when $\sum_{n=1}^{\infty} (b_n - a_n) = 1$— φ is a counterexample to Harnack's extension of the Fundamental Theorem III (3.1).

According to Cantor, Ludwig Scheeffer (1859–85) had already discovered a similar counterexample by an entirely different line of reasoning. In conjunction with his study of the concept of curve length Scheeffer [1884a: 61–62, 67–68] considered the following function, which represents a rectifiable curve:[10] Let $w_1, w_2, \cdots, w_r, \cdots$ and $s_1, s_2, \cdots, s_r, \cdots$ denote two sequences of real numbers such that the s_r are positive and the series $\sum_{r=1}^{\infty} s_r$ converges. Define a function f by

$$f(x) = \sum_{-\infty}^{x} s_r,$$

where the right-hand side denotes a summation over all s_r such that w_r is less than x. At each w_r, f has a jump discontinuity, since

$$f(w_r+) - f(w_r-) = f(w_r+) - f(w_r) = s_r.$$

Suppose now that the w_r form a dense subset of (a,b), and consider

10. Scheeffer's work on rectifiable curves is discussed more fully in 3.4.

$f([a,b])$. Since f is strictly increasing and has jump discontinuities, $f([a,b]) = [f(a), f(b)] - \bigcup\limits_{r=1}^{\infty} (f(w_r), f(w_r+)]$. An "inverse function" g can be defined on $[f(a), f(b)]$ by $g(y) = f^{-1}(y)$ if y is in $f([a,b])$ and $g(y) = w_r$ if y is in $(f(w_r), f(w_r+)]$, for $r = 1, 2, 3, \cdots$. Then $g'(y) = 0$ for y in $\bigcup\limits_{r=1}^{\infty} (f(w_r), f(w_r+)]$. The sum of the lengths of the intervals $(f(w_r), f(w_r+))$ is $\sum\limits_{r=1}^{\infty} s_r = f(b) - f(a)$ so that $f([a, b])$ is a discrete subset of $[f(a), f(b)]$. Thus $g'(y) = 0$ except on a discrete set.

Harnack [1884b: 287n; 1884c: 225–26] had also come to recognize the existence of functions of this sort, which represent counterexamples to his generalized version of the Fundamental Theorem III (3.1). The generalized version had served as the basis for his extension of Riemann's integral to functions f that become unbounded in the neighborhood of points forming a discrete set, U_f [Harnack 1882a: 245–46]. (On U_f, f may assume infinite values or be undefined.) The first such attempt to extend the integral of Riemann to unbounded functions had been made by Paul du Bois-Reymond; he [1875a: 37] observed that if U_f is a first species set, then the usual procedure of defining improper integrals when U_f is a finite set can be carried over by induction to the case in which $U_f^{(n)}$ is finite. (Here and throughout this discussion it is always assumed that the functions under consideration are Riemann-integrable in any interval not meeting U_f.) Harnack's definition was of a different sort: If U_f is discrete, f is defined to be integrable if there exists a continuous function F defined on $[a,b]$ such that $F(a) = 0$ and $F'(x) = f(x)$ "in general." That is, for every positive ϵ, the set of points x at which

$$\left| F'(x) - f(x) \right| \geq \epsilon$$

is discrete. (The set of points x for which $F'(x)$ does not exist is also assumed to be a discrete set.) Then $\int_a^b f$ is defined to be $F(b) - F(a)$. Once Harnack's version of the Fundamental Theorem III was discredited, it became clear that F, and therefore $\int_a^b f$, need not be uniquely defined; for if F satisfies the conditions that $F(a) = 0$ and $F'(x) = f(x)$ "in general," then so does $G(x) = F(x) + g(x) - g(a)$, where g is any nonconstant function such that $g'(x) = 0$ except on a discrete set.

Both du Bois-Reymond and Harnack had been motivated to extend the integral to unbounded functions by the theory of trigonometric series. By extending the concept of the integral, they hoped to extend the validity of Fourier's assumption that every trigonometric series is a Fourier series. Although du Bois-Reymond [1875b] was able to handle the case in which U_f is of the first species, Harnack's results were vitiated by his unsuccessful generalization of the integral. One of du Bois-Reymond's former students, Otto Hölder (1859–1937), was induced by Harnack to consider the problems involved when U_f is not of the first species [Hölder 1884: 191 ff]. The first step was, of course, to obtain a valid generalization of the integral. Hölder showed that a definition along the lines of Harnack's could be given provided U_f is required to be countable. Specifically, Hölder considered a function f defined on $[-\pi,\pi]$ which is Riemann-integrable in the neighborhood of every point except those in a countable set E. Then f is integrable if there exists a continuous function F such that for every fixed t not in E the function of $x, F(x) - \int_t^x f$, remains constant in any neighborhood of t on which f is Riemann-integrable; by definition,

$$\int_{-\pi}^{\pi} f = F(\pi) - F(-\pi).$$

Hölder's proof that F must be unique draws upon the results of Cantor's research. Suppose there exists another function $F_1(x)$ which satisfies the conditions. Then $g(x) = F(x) - F_1(x)$ is constant in a neighborhood of every t not in E. Obviously with the counterexamples of Cantor and Scheeffer in mind, Hölder considered the nature of the set R, the set of points x such that g is not constant in a neighborhood of x. It is not difficult to see that R is a perfect set and hence, by Cantor's results, either empty or uncountable. Since R is a subset of E, however, R must be empty. On the basis of this definition of the integral, Hölder was then able to show that if

$$f(x) = \tfrac{1}{2}a_0 + \sum_{n=1}^{\infty} (a_n \cos nx + b_n \sin nx),$$

and if the functions $f(x)\cos nx$ and $f(x)\sin nx$ are integrable, then a_n and b_n are the Fourier coefficients of f. (It does not follow that these functions possess Hölder integrals if f does.)

The definitions of the integral proposed by Harnack and Hölder are not constructive: they do not provide a method for defining the function F. They also take as fundamental the notion of the indefinite integral. Later Harnack [1883: 324–26] proposed another definition, which is more in line with the customary, constructive procedure for defining improper integrals and at the same time is applicable when U_f

is any discrete set: Let $I_1, I_2, \cdots, I_n$ denote a finite number of intervals such that U_f is contained in U, where $U = \bigcup\limits_{k=1}^{n} I_k$. Let f_U denote the function such that $f_U(x) = f(x)$ for x not in U and $f_U(x) = 0$ for x in U. Then f_U is Riemann-integrable. The function f is defined to be integrable if $\int_a^b f_U$ approaches a limit as $\sum\limits_{k=1}^{n} L(I_k)$ approaches 0.

Harnack was ambiguous as to whether the covers $U = \bigcup\limits_{k=1}^{n} I_k$ are allowed to include intervals I_k that are superfluous (i.e., that contain no points of U_f). This distinction is an important one for the nature of the resultant definition.

In a paper written after Hölder's, Harnack [1884c: 220 ff] returned to this definition. If f has a Harnack integral on $[a,b]$, then the function $F(x) = \int_a^x f$ is continuous, and f together with F fulfill the conditions of Hölder's definition provided that U_f is countable. It thus follows, as Harnack observed, that when U_f is countable, Harnack integrability implies Hölder integrability. Harnack sought to show that the converse is also true, and, therefore that his integral generalizes Hölder's definition; but this is not the case (see 5.1). Harnack's attempted proof is nevertheless important historically because it called attention to the property that has since been termed absolute continuity. He began with the observation that if f is Riemann-integrable, then $F(x) = \int_a^x f$ satisfies the following condition:

(12)   *If (a_i, b_i), $i = 1, 2, \cdots, n$, are nonoverlapping intervals (n is not fixed), then $\sum\limits_{i=1}^{n} [F(b_i) - F(a_i)]$ converges to 0 with $\sum\limits_{i=1}^{n} (b_i - a_i)$.*

This condition, although seemingly less restrictive than that of absolute continuity, is actually equivalent to it.[11]

The importance of condition (12) and the definition of the Harnack integral itself were probably suggested to Harnack by a passage in

11. The usual definition requires that $\sum\limits_{i=1}^{\infty} |F(b_i) - F(a_i)|$ converge to 0 with $\sum\limits_{i=1}^{\infty} (b_i - a_i)$. By considering separately the sums of positive and negative terms $F(b_i) - F(a_i)$, its equivalence with (12) is easily proved.

Ulisse Dini's book *Serie di Fourier e altre rappresentazioni analitiche delle funzioni di una variabile reale* [1880]. In order to extend the second mean value theorem for integrals to improper integrals as defined by du Bois-Reymond, Dini proved the following lemma: The first species set U_f "can be enclosed in a finite number of intervals so small that the absolute value of the sum of the integrals extended over these intervals . . . is less than an arbitrarily small number σ" [1880: 22]. Harnack refers to this passage in connection with his introduction of the Harnack integral, which was motivated by his desire to remove the condition that U_f be a first species set while at the same time maintaining the validity of Dini's lemma and, hence, of the useful second mean value theorem.

The motivation for introducing (12) is clear: if f is Harnack-integrable and if $U = \bigcup_{i=1}^{n} (a_i, b_i)$ is an interval cover of U_f, then the Harnack integral $F(x) = \int_{a}^{x} f$ has the property that

$$\sum_{i=1}^{n} [F(b_i) - F(a_i)] = \int_{a}^{b} f - \int_{a}^{b} f_U$$

converges to 0 with $\sum_{i=1}^{n} (b_i - a_i)$. Now suppose that f has an indefinite Hölder integral F. Then the definition of F implies that

$$\int_{a}^{b} f_U = [F(a_1) - F(a)] + [F(a_2) - F(b_1)] + \cdots + [F(b) - F(a_n)]$$

$$= F(b) - F(a) - \sum_{i=1}^{n} [F(b_i) - F(a_i)],$$

where the notation is chosen so that

$$a \le a_1 < b_1 < a_2 < \cdots < b_{n-1} < a_n < b_n \le b.$$

Thus, *if* F satisfies (12), it will follow—by letting $\sum_{i=1}^{n} (b_i - a_i)$ converge to 0—that the Harnack integral of f over $[a,b]$ is precisely $F(b) - F(a)$. The trouble is that F need not satisfy (12), although Harnack thought he could prove that F satisfies (12) from the fact that U_f is reducible.

However, Harnack did recognize the difficulties that arise when U_f is not reducible: Unless F is an "integral function"—that is, unless F satisfies (12)—then (1) F can satisfy the other parts of Hölder's definition and still not differ from the Harnack integral $\int_{a}^{x} f(t)dt$ by a constant, and (2) the Fundamental Theorem II is not valid for Harnack integrals. The reason for both difficulties is the same: There exist continuous nondecreasing functions with dense intervals of invariability for which the set of points outside these intervals is discrete. If F denotes such a function, then D^+F, for example, possesses a Harnack integral, since D^+F is zero off a discrete set. But

$$\int_a^x D^+F = 0 \neq F(x) - F(a),$$

and the Fundamental Theorem II fails to hold.

Thus, while the theory of sets suggested ways to extend the Riemann integral to unbounded functions, it also revealed the existence of a disturbing class of functions. These functions showed that the difficulties encountered by Dini and Volterra in connection with the Fundamental Theorem II are compounded when Riemann's definition is extended to unbounded functions, since even the integrability of a derivative of an unbounded function is insufficient to guarantee the validity of Fundamental Theorem II. These functions, however, also focused attention upon the distinctive property of continuous functions which are integrals (see 5.1).

3.4 Curve Length and the Integral

A deeper familiarity with infinite sets of points had revealed that the fundamental formula $\int_a^b f' = f(b) - f(a)$ could no longer be generally maintained for Riemann's definition of the integral and its extensions to unbounded functions. The theory of sets also stimulated an interest in the classical integral formula for the length of a curve $y = f(x)$, viz.:

$$(13) \qquad L = \int_a^b [1 + (f')^2]^{1/2}.$$

The interest led to the discovery that the classical formula is not generally valid. Although there was some difference of opinion, it was generally felt that the integral concept and in particular (13) are not relevant to the general theory of curve rectification, a theory which had evolved within the framework of the above-mentioned discovery. These developments are especially important because Lebesgue, who was familiar with them, was able to see that his generalization of the integral was just what was needed to reinstate the earlier belief in the close connection between the concepts of curve length and the integral.

Prior to the nineteenth century, little if any attention had been paid to the questions of the definition and of the existence of the length of a curve, just as little attention had been paid to the analogous questions for the integral of a function. Curves of the form $y = f(x)$ were assumed to have a length that could be expressed by (13), and also regarded as the limiting value of the lengths of inscribed polygonal lines. Paul du Bois-Reymond was the first mathematician to consider the concept of curve length from the standpoint of Dirichlet's and Riemann's general concept of a function and in relation to the nascent theory of sets. The necessity for such considerations arose in connection with his study [1879a] of the basic operations of the calculus of variations as applied to the problem of determining the curve of shortest length between two points. The problem, to be both well-defined and general, necessitated giving as extensive a definition as possible of a rectifiable curve $y = f(x)$. For a curve which is not a straight line, he argued, the concept of its length involves a passage to the limit; it is an analytical notion. To his way of thinking, this implied that curve length must be defined in terms of the integral; this preconception determined the manner in which he sought to extend the notion of a rectifiable curve.

Specifically, du Bois-Reymond considered any curve represented by a function f defined on $[a,b]$ such that f is continuous except for jump discontinuities and that P, the set of x at which f has a jump discontinuity, is a first species set. He further assumed that the function possesses a bounded, integrable derivative on any interval not meeting P. If, for example, P consists of a finite number of points x_i, $1 \leq i \leq n$, where

$$a = x_0 < x_1 < x_2 < \cdots < x_n < x_{n+1} = b,$$

then the length L of the curve is defined to be

$$L = \sum_{i=1}^{n} |f(x_i+) - f(x_i-)| + \sum_{i=0}^{n} \int_{x_i}^{x_{i+1}} [1 + (f')^2]^{1/2}.$$

The definition of L is then extended by induction on the type n of P under the additional condition that the series

$$\sum_{x \in P} |f(x+) - f(x-)|$$

converge. Du Bois-Reymond was convinced that he had accorded the greatest generality possible to the concept of rectifiability. Indeed the existence of points of discontinuity densely distributed in some interval, however small, appeared inimical to this concept, since then there would be, he claimed, no derivative [1879a: 287]. This insistence upon the general existence of $f'(x)$ reflects, of course, du Bois-Reymond's initial assumption that rectifiability is equivalent to the existence, admittedly in a somewhat extended sense, of the integral in (13). It should also be noted that du Bois-Reymond's reasoning tacitly assumes that if the discontinuity points do not form a first species set, then they will be dense in some interval; apparently in 1879 he was still confusing nowhere dense and first species sets.

Du Bois-Reymond's paper thus raised the question of the rectifiability of curves determined by these singular functions—functions that had come into importance with the development of the theory of integration but had been traditionally, albeit unconsciously, precluded from discussions of curve rectification. Two years later in an appendix to a paper on Bolzano's significance in the development of mathematical analysis, the subject was taken up by Otto Stolz [1881]. Bolzano had regarded the length of a continuous curve simply as a magnitude, which he did not further define. Stolz was greatly impressed by the treatment of this concept which he found in the multivolume *Des méthods dans les sciences de raisonnement* (1865–73), a work written by J. M. C. Duhamel (1797–1872) and devoted to the principles of mathematics.

In the second volume, published in 1866, Duhamel defined the length of a curve to be the limit of the lengths of polygons whose vertices coincide with points of the curve (including the endpoints) as the length of their sides tends to zero. The idea behind the definition was, of course, hardly original with Duhamel. But Duhamel saw the necessity of *defining* curve length as this limiting value,[12] and he realized that "this definition will have a precise sense only if it is proved that this limit exists and is unique, no matter what the rule is according to which the sides of the polygon tend to zero . . . " [1866: 412]. The possibility of nonrectifiable curves, however, was not a real one for him: an "arbitrary curve" was assumed to be endowed with enough properties to make a proof possible.

Although Stolz accepted Duhamel's definition of curve length, he followed du Bois-Reymond in believing that if conditions sufficient for the existence of the integral in (13) were determined, the class of rectifiable curves could be extended beyond its customary limits. Du Bois-Reymond's generalization had been in the direction of dropping the assumption that $f(x)$ be everywhere continuous. Stolz restricted himself to continuous functions f such that f' is integrable in the sense of du Bois-Reymond and Dini. (That is, in the notation of 3.3, $U_{f'}$ is assumed to be of the first species.) He was able to prove that the necessary and sufficient condition for the rectifiability of $y = f(x)$ is

12. A similar definition of curve length had been given by E. H. Dirksen [1833].

the existence of the improper integral $\int_a^b |f'|$ in the sense of du Bois-Reymond; furthermore $\int_a^b |f'|$ is coexistent with $\int_a^b [1+(f')^2]^{1/2}$, which always gives the length of the curve.

Thus, granted such restrictions on f and f', it can be said that Stolz maintained the harmony between the concepts of integral and curve length. But it was soon pointed out by Ludwig Scheeffer (1859–85) that the acceptance of Duhamel's definition of curve length and rectifiability is really not compatible with the traditional approach to curve length based on the integral [1884a]. According to Scheeffer, his researches on the rectifiability of curves were motivated by his study of a "remarkable function" that is continuous but whose derivative "possesses the same non-zero oscillation in every arbitrarily small interval so that the definite integral $\int_{x_0}^{x_1} [1+f'(x)^2]^{1/2}dx$ does not have a meaning on the basis of Riemann's definition" [1884a: 49]. Nevertheless, on the basis of "geometrical considerations" the curve $y=f(x)$ has a definite length. Thus, he argued, the notion of curve length cannot be made to depend on the existence or nonexistence of the integral.

Scheeffer's objective was to determine conditions for the rectifiability of a completely arbitrary curve $y=f(x)$, rectifiability being understood in Duhamel's sense. He succeeded in obtaining several sets of conditions for rectifiability, although he failed to recognize that the crucial condition is that the function f be representable as the difference of increasing functions. But his results did suffice to show, for example, that the function $f(x) = \sum_{-\infty}^{x} s_r$ (discussed in 3.3) represents a rectifiable curve even though the points of discontinuity, the w_r, can form a dense set. Thus, du Bois-Reymond's stipulation that $y=f(x)$ be continuous except on a nowhere dense set "turns out to be inessential with our definition of length . . . " [1884a: 54n]. Scheeffer also considered the function

$$f(x) = \sum_{n=1}^{\infty} c_n(x - w_n)^{1/3},$$

where the c_n are positive, $\sum_{n=1}^{\infty} c_n$ converges, the w_n are dense in $(-\infty, \infty)$, and the c_n are chosen so that the above series converges uniformly. This example had been considered by Weierstrass but was published first by Cantor [1882a], who observed that the method of construction suggests a more satisfactory principle of condensation of singularities than that proposed by Hankel. The function is strictly increasing and continuous, although the integral $\int_a^b [1+(f')^2]^{1/2}$ "by which the length is customarily expressed is completely meaningless in this case," since $f'(w_r) = +\infty$ at each w_r [Scheeffer 1884a: 67]. It was probably this function that Scheeffer referred to as providing the stimulus for his work on rectifiable curves.[13] (Under suitable conditions on the c_n, this function also has an inverse that is differentiable and possesses a bounded derivative but is not Riemann-integrable; see 4.3.)

It will be recalled that Scheeffer had discovered the unusual properties of $g(x)$, the "inverse" of $f(x) = \sum_{-\infty}^{x} s_r$. Both he and Harnack [1884c: 230] pointed out that for functions such as g the integral formula $\int_a^b [1+(g')^2]^{1/2}$—interpreted as an improper Harnack integral—gives the erroneous value of $b-a$ for the arc length of $y = g(x)$.

Thus, by theorem and example Scheeffer supported his thesis that the existence of tangents and the integrability of $[1+(f')^2]^{1/2}$ are not essential for or relevant to the rectifiability of the curve $y = f(x)$ and the determination of its length. His paper drew an immediate response from du Bois-Reymond [1885], who attempted to defend his original position by arguing that the concept of curve length—although it is a purely analytical extension of the concept of a straight line—should nevertheless remain within the customary bounds of geometry. In geometry, as well as in mechanics and the calculus of variations, he argued, the existence of a tangent is taken as a property of curves; it seemed questionable whether one could speak of the geometrical equivalent of a nondifferentiable function—a rather curious statement from one who had himself proposed to extend the concept of curve length to functions that are discontinuous on a first species set.

Closely connected to the rectification of curves is the notion of a function of bounded variation. In the course of his studies on the fundamental formula behind Dirichlet's theorem on the convergence of Fourier series, du Bois-Reymond [1880a: 32–35] called attention to the fact that, since the proof of Dirichlet's theorem depends on the

13. For this function, the oscillation of $f'(x)$ in every interval is $+\infty$, in agreement with Scheeffer's description. It would be impossible to have an example of a function $f(x)$ such that the oscillation of $f'(x)$ is a finite nonzero constant value in every interval, for then $f'(x)$ would be totally discontinuous, a situation that is impossible (as shown by a theorem of Baire's).

monotonicity of the function under consideration, the theorem remains valid for any function expressible as the difference of two nondecreasing functions. He was, however, unable to completely determine the nature of these functions and limited himself to the observation that if f has integrable Dini derivatives Df, then f can be represented as the difference of nondecreasing functions: $f(x) = f(a) + \int_a^x g - \int_a^x h$, where $g(x) = \max(Df(x), 0)$ and $h(x) = \max(-Df(x), 0)$. Dini had, of course, called attention to this property for the larger class of first species functions (2.3).

The same observation concerning Dirichlet's theorem was made two years later [1881] by Camille Jordan (1838–1922). But, by introducing the concept of functions of bounded variation, Jordan succeeded in completely characterizing functions representable as the difference of nondecreasing functions. Functions of bounded variation were defined in the following manner: Let $f(t)$ be defined on $[a,b]$, and let x belong to $(a,b]$. Consider a partition $a = t_0 < t_1 < \cdots < t_n = x$ of $[a,x]$ and set $y_i = f(t_i)$, $i = 0, 1, \cdots, n$. The points (t_i, y_i) determine a polygon. The sums $\sum' (y_i - y_{i-1})$ and $-\sum'' (y_i - y_{i-1})$ are the positive and negative variation of the polygon, where $\sum'$ and $\sum''$ signify summation over all terms $(y_i - y_{i-1})$ that are positive and negative, respectively. If, as the polygon varies, the corresponding positive and negative variations have finite least upper bounds $P(x)$ and $N(x)$ for each x in $(a,b]$, then f is defined to be of bounded variation (*à oscillation limitée*) on $[a,b]$; $P(x)$, $N(x)$, and $T(x) = N(x) + P(x)$ are called the positive, negative, and total variation, respectively, of f on $[a,x]$.

It is clear that Jordan could have defined functions of bounded variation by requiring simply that $T(x)$, the supremum of $\sum_{i=1}^{n} \left| y_i - y_{i-1} \right|$, be finite. By considering the positive and negative terms separately, however, the connection between functions of bounded variation and monotonic functions comes to light:

$$f(x) - f(a) = \sum_{i=1}^{n-1} (y_i - y_{i-1}) = \sum' (y_i - y_{i-1}) - [- \sum'' (y_i - y_{i-1})],$$

so, in the limit:

$$f(x) - f(a) = P(x) - N(x).$$

It is clear from their definitions that P and N are nondecreasing functions; thus, f can be represented as the difference of nondecreasing functions. Conversely, if $f(x) = f_1(x) - f_2(x)$ and f_1 and f_2 are nondecreasing, then the positive and negative variations of the polygon are less than or equal to $f_1(x) - f_1(a)$ and $f_2(x) - f_2(a)$, respectively; f is therefore of bounded variation. In this manner Jordan obtained the fundamental theorem that f is of bounded variation if and only if it can be expressed as the difference of nondecreasing functions.[14]

The properties of functions of bounded variation became widely known because they were discussed by Jordan in a note appended to the third volume of his *Cours d'analyse* (1887). Here the close connection between the concepts of a rectifiable curve and of functions of bounded variation, already suggested by his terminology in [1881], is made explicit. Jordan also made the observation that indefinite Riemann integrals $F(x) = \displaystyle\int_a^x f$ are of bounded variation—a fact which plays an important role in Lebesgue's work.

In the second edition of his *Cours*, the discussion of functions of bounded variation and rectifiable curves was incorporated into the text itself (Vol I, 1893). At this time Jordan added the theorem that if $x = x(t)$, $y = y(t)$ represents a rectifiable curve C for which $x'(t)$ and $y'(t)$ exist and are continuous, then $s'(t) = [x'(t)^2 + y'(t)^2]^{1/2}$, where $s(t)$ denotes the arc length between t_0 and t. An immediate corollary would be that the length of C is given by the classical integral formula, but, curiously enough, Jordan chose not to make this remark. In fact, nowhere in the 1887 note or in the discussion of rectifiable curves in the second edition did Jordan present the integral formula. This omission is perhaps indicative of his acceptance of Scheeffer's opinion that the theory of integration is of no significance for the general theory of curve rectification. As E. Study wrote in a paper on functions of bounded variation: "We therefore cannot help but consider the application of the concept of length to curves without tangents as completely justified; and we must accept the view expressed by Scheeffer that the usual definition of arc length by means of a definite integral contains a groundless restriction" [1896: 316]. And, according to Study, Scheeffer's view was shared by Jordan as well.

In his *Grundzüge der Differential- und Integralrechnung* [1893–99: II] and in a subsequent paper [1902], Otto Stolz returned to the problem of the validity of the integral formula for arc length, but obtained no advance over his earlier result. Success in this direction required a more flexible definition of the integral and the genius of Lebesgue.

14. Independently of Jordan and Scheeffer, G. Ascoli discovered that $y = f(x)$ is always rectifiable when f is monotonic, regardless of whether f is differentiable or continuous [1883; 1884], but he failed to go beyond this observation.

4

The End of
the Century: A Period
of Transition

For the theory of integration, the last decade of the nineteenth century represents a period of transition from old ideas to new. The notion of measurability, introduced for somewhat different reasons by both Jordan and Giuseppe Peano, formed the crucial link of the transition. It made possible the reformulation of Riemann's definitions of the integral and integrability within a measure-theoretic context, and it is in this formulation that their close relation to Lebesgue's definition first becomes apparent. Furthermore, no sooner had that occurred than Émile Borel postulated a radically different notion of measurability, which was destined to lead the way to a new theory of integration. The manner in which Borel's ideas were treated in Arthur Schoenflies' widely read report on the theory of sets, however, suggests that at first they appeared too radical for many mathematicians. Also during the late nineteenth century—through the discovery of further classes of functions with bounded and nonintegrable derivatives, and through researches on the term-by-term integration of infinite series of functions—the inability of Riemann's theory to handle limiting processes satisfactorily became more apparent.

4.1 THE INTRODUCTION OF THE CONCEPT OF MEASURABILITY

The concept of measure adopted by Stolz, Cantor, and Harnack was, for most mathematicians with the exception of Weierstrass, dissociated from the concept of the definite integral; that is, it was assumed that the region E bounded by the graph of a function always has an outer content regardless of whether the function possesses an integral. A different attitude toward the relationship between these concepts was taken by Giussepe Peano (1858–1932). Peano [1883] showed that proofs in the theory of integration could be simplified by recognizing that the upper and lower integrals, $\overline{\int_a^b} f$ and $\underline{\int_a^b} f$, can be defined as the greatest lower bound and the least upper bound of the upper and lower Riemann sums of f. Also, Peano was critical of treatments of the integral in which its definition and existence are based upon the concept of area—not because of the "geometrical" orientation but because the notion of area was never given a precise, sufficiently general definition.

For a region in the plane "of simple form" Peano gave what he considered the most natural definition of its area: Consider the two classes of polygons that contain and are contained within the given region. The areas of the polygons in the first class have a greatest lower bound; in the second class, a least upper bound. If these limits coincide, their common value is by definition the area of the region. On the other hand, if they differ, "the concept of area would not apply in this case. Therefore, in order to speak of the area of a figure, it is first necessary to verify that these two limits are equal, which is nothing but the preceding integrability condition" [1883: 446]. This last remark makes it clear that Peano was led to his conception of area—more precisely, his conception of the criterion for the existence of an area—by stressing the analogy with the criterion of Riemann integrability (R_1), whereas the definition of outer content had been motivated by Riemann's condition (R_2).

The ideas expressed in passing in 1883 received a detailed, completely general treatment in the chapter "Geometrical Magnitudes" of Peano's *Applicazioni geometriche del calcolo infinitesimale* [1887]. Cantor's researches on point sets seem to have influenced Peano to present his concept of area in the context of completely arbitrary sets.[1] Peano began by generalizing and at the same sharpening the intuitive conceptions of interior, boundary, and exterior points associated with regions bounded by curves [1887: 152ff]. In Euclidean space of one, two, or three dimensions, a point p is an interior point of the set A if there exists a positive δ such that every x whose distance from p is less than δ belongs to A; p is exterior to A if there is a positive δ such that all

---

1. It is clear that Peano was familiar with Cantor's work, especially since he used the term "closed set" in the sense defined by Cantor.

points x whose distance from p is less than δ do not belong to A; and p is a boundary point if it is neither an interior nor an exterior point. Peano noted that the boundary of a set could be highly unusual as in the case of the set of rational numbers between 0 and 1, since every point of [0,1] is a boundary point of that set.

The definitions of the measure of a set in one-, two-, and three-dimensional space were handled separately by Peano. For A, a subset of two-dimensional space, he defined the "inner area" of A as the least upper bound of the areas of all polygonal regions contained within A and the "outer area" as the greatest lower bound of all polygonal regions containing A. If these two numbers coincide, the common value is the area of A; otherwise—as Peano put it—A does not have an area comparable with the area of a polygon. Peano recognized the relationship (in modern notation) $c_e(A) = c_i(A) + c_e(\partial A)$ and its implication that A has an area if and only if $c_e(\partial A) = 0$.

Peano pointed out that the set function $c(A)$ is an example of what he referred to as a distributive function, that is, a finitely additive set function. His chapter on geometrical magnitudes is largely devoted to the study and application of distributive functions and, in particular, to a theory of differentiation and integration in the context of these functions. The theory is surprisingly elegant and abstract for a work of 1887 and strikingly modern in its approach. Indeed, by making the notion of a set function fundamental to the theory of differentiation and integration—an idea that he borrowed from Cauchy[2]—Peano anticipated the viewpoint that was to produce the Lebesgue-Stieltjes integral. Apparently, however, Peano's work was not influential in bringing about this viewpoint.

In *Applicazioni geometriche* Peano again pointed out the close connection between the concepts of measure and of the integral: If $f(x)$ is nonnegative for x in $[a,b]$, then

$$\underline{\int_a^b} f = c_i(E) \quad \text{and} \quad \overline{\int_a^b} f = c_e(E),$$

2. Cauchy [1841] defined two magnitudes to be "coexistent" if they "exist together and vary simultaneously" and in such a manner that the "elements" of the one exist and vary or vanish at the same time as the "elements" of the other. Thus, for example, the mass and volume of a body are coexistent. Cauchy then defined the differential quotient of two coexistent magnitudes A and B as the limit of the ratio a/b as $b \rightarrow 0$, where a and b are corresponding elements of A and B. The general treatment is extremely vague, but it is clear from his examples that A and B are essentially set functions and that Peano's definition of differentiation makes precise what Cauchy meant by differentiation. Cauchy, however, did not treat the theory of integration from this standpoint.

where E denotes the region bounded by the graph of f [1887: 194–95]. Thus f is integrable if and only if E is "measurable," i.e., $c_i(E) = c_e(E)$.

Although the notion of measurability is clearly basic to Peano's treatment of the measure of sets, it was Camille Jordan who, five years later, explicitly introduced and established the importance of the concept of a measurable set. Jordan's motivation came from the theory of multiple integrals. Here the property of measurability arises quite naturally as soon as one attempts to extend the theory of, for example, the double integral $\int_E f(x,y)dE$ to an arbitrary set E. Jordan was not the first one to recognize this. During the period 1850–92 there were a number of works on double integrals which explicitly refer to and make use of the property of measurability.

Jordan's contribution to measure theory is best appreciated by an understanding of the earlier treatments of this property. Typical examples of the manner in which this occurred would be as follows: Suppose the plane is partitioned into rectangles R_{ij} of dimensions Δx_i and Δy_j by lines parallel to the coordinate axes, thus inducing a partition of E into subsets E_{ij}. Most of the E_{ij} will be rectangles, but some will be irregular and contain part of the boundary of E. By analogy with the one-dimensional case, $\int_E f(x,y)dE$ can be defined as the limit of the Cauchy sums

$$(1) \qquad \sum_{ij} f(x_i,y_j)a(E_{ij})$$

as the dimensions of the R_{ij} approach zero. (Each (x_i,y_j) is in E_{ij}.) The meaning of the area of E_{ij}, denoted here by $a(E_{ij})$, was left unquestioned. In fact, no symbolic differentiation was made between the set E_{ij} and its area. In more careful treatments, however, this difficulty was avoided by defining the double integral over E to be the limit of the sums

$$(2) \qquad \sum_{ij}' f(x_i,y_j)a(R_{ij}),$$

where $\sum'$ denotes summation either over all R_{ij} contained entirely in E or over all R_{ij} that meet E. For such a definition to be free from an element of arbitrariness, it must be assumed that the sum of the areas of those R_{ij} meeting the boundary of E approaches zero with the dimensions of the R_{ij}. In other words, it must be assumed that E is measurable. Even when this amount of care was not taken in defining the double integral, it was found convenient, when showing that a double integral has the same value as iterated single integrals, to assume that sums of the form (2) could be used as well as those of form (1). This assumption, because $a(R_{ij}) = \Delta x_i \Delta y_j$, allows (2) to be arranged, for example, as

$$\sum_i \left[\sum_j f(x_i, y_j) \Delta y_j \right] \Delta x_i.$$

The use of sums (2) rather than sums (1) is justified by assuming that the sum of the nonrectangular E_{ij} converges to zero with the dimensions of the R_{ij}; so again the idea of measurability enters the argument.

Among those who thus made use of the property of measurability there were two distinct levels of apprehension of its significance. Some simply took for granted that measurability was an obvious property of E—an assumption probably due to the fact that E was traditionally regarded as a region enclosed by curves, which in turn were assumed to have zero area. Typical of those holding this viewpoint was Karl J. Thomae, who appears to have been the first to extend Riemann's theory of integration to functions of two variables [1876]. The extension is fairly routine except at one point. A number of mathematicians, including Thomae, had discovered and emphasized the fact that the upper and lower Riemann sums of an arbitrary bounded function always approach limits as the norm of the partitions approaches zero (2.2). When the proof of this fact is carried out for functions of two variables, the measurability of E as well as of the E_{ij} enters into the argument. Thus Thomae considered two partitions, τ and t, of a set T on which a bounded function $f(x,y)$ is defined; then, "the subdivision τ can be carried so far that the area of the sum of all those τ which cross over some part of the boundary of t or only touch it becomes less than $\sigma:M$, where M is the upper limit of the absolute value of the values of $f(x,y)$ in T; for this area can clearly be made arbitrarily small" [1876: 226]. Similarly, Jordan himself in the first edition of his *Cours d'analyse* (Vol 2, 1883) justified the transition from the sums (1) to those of form (2) with the comment that the nonrectangular E_{ij} "form an infinitely thin rim about the contour" [1882–87: II, 127].[3]

On the other hand, some mathematicians definitely recognized that an additional assumption was involved. Harnack, for example, added the following footnote to his discussion of double integrals: "It is assumed when speaking of a boundary curve that the boundary points of the plane region are of such a nature that they can be enclosed in a

---

3.   This type of treatment is also found in Dirichlet's lectures (1854 and later years) on definite integrals [Dirichlet 1904: 224–25; Meyer 1871: 438–39] and in the treatises of Lipschitz [1880: 539], Boussinesq [1890: 90], Hermite [1891: 37ff], and Gilbert [1892: Sec. 337].

plane domain whose magnitude is arbitrarily small. The same holds also for the curves by means of which the region P is partitioned" [1884–85: 278]. Although Harnack limited his discussion to sets determined by boundary curves, his remark indicates that he realized that in the definition of $\int_E f(x,y)dE$ not only E but also the subsets into which it is partitioned must be measurable. The property of measurability was also postulated by a young Italian mathematician, Rodolfo Bettazzi [1884: 147], within the context of arbitrary sets, although his paper was apparently not widely read. Other mathematicians preferred to state conditions on E or, more precisely, on the curve bounding E. They then deduced from these conditions that E has the property of being measurable. For example, Cesare Arzelà (1847–1912) included the hypothesis that E is bounded by a simple closed continuous curve which is rectifiable [1891]. Émile Picard (1856–1941) assumed that lines parallel to the coordinate axes meet the boundary curve of E in at most N points (N a fixed integer), although he realized the lack of necessity of this condition for the measurability of E [1891: I, 93ff]. Arzelà and Picard were probably influenced to pay more than the usual attention to the boundary curve of E because of Peano's discovery [1890] of the existence of continuous curves that pass through every point of a square. Peano stressed the fact that further assumptions must be imposed on a continuous curve if it is to be enclosed in a region of arbitrarily small area.

Before Jordan's work with measurable sets, discussions of the relation between a double integral and iterated single integrals also involved the concept of measurability. For the sake of brevity we shall refer to all theorems positing the identity of double and iterated single integrals as "Fubini's Theorem," regardless of the degree of generality entering into the hypotheses.

Early in the nineteenth century Cauchy had pointed out that the repeated integrals

$$\int_0^1 dy \left[\int_0^1 f(x,y)dx \right] \quad \text{and} \quad \int_0^1 dx \left[\int_0^1 f(x,y)dy \right]$$

need not be equal when f is unbounded [1827a: 394]. Interest in this matter was revived by Thomae [1878] when he gave a simple example of a bounded function $f(x,y)$ for which the second integral exists while the first is meaningless because, except for one value of y, the function defined by the correspondence $x \rightarrow f(x,y)$ is not Riemann-integrable.

In the examples of Cauchy and Thomae, the double integral of

$f(x,y)$ did not exist. But du Bois-Reymond [1883a] showed that even when $f(x,y)$ is integrable as a function of two variables, the functions $x \rightarrow f(x,y)$ and $y \rightarrow f(x,y)$ need not be integrable for all values of y and x, respectively. To illustrate this he considered the following function: $f(x,y) = 1/2^p$ whenever (x,y) is of the form $([2n+1]/2^p, [2m+1]/2^q)$ for some positive integers m, n, p, q; $f(x,y) = 0$ otherwise. This function is integrable over the rectangle R determined by $0 \le x \le 1$ and $0 \le y \le 1$, although the function $y \rightarrow f(x,y)$ is totally discontinuous when $x = (2n+1)/2^p$. Thus, for a dense set of values of x—a set of outer content 1—the integral $\int_0^1 f(x,y)dy$ fails to exist. Nevertheless, du Bois-Reymond was able to show that Fubini's Theorem could still be maintained in a modified form, which can be stated as follows for upper integrals: If $f(x,y)$ is integrable on R, then the functions

$$y \rightarrow \overline{\int}_0^1 f(x,y)dx \quad \text{and} \quad x \rightarrow \overline{\int}_0^1 f(x,y)dy \quad \text{are likewise integrable and}$$

$$\int_R f(x,y)dR = \int_0^1 dy \left[\overline{\int}_0^1 f(x,y)dx \right] = \int_0^1 dx \left[\overline{\int}_0^1 f(x,y)dy \right].$$

But du Bois-Reymond did not consider the validity—or meaning—of Fubini's Theorem when E is completely general. In this case, for example, the inner integral occurring in the iterated integral $\int d\eta \left[\int f(x,\eta)dx \right]$ would be integrated over G_η, the set of points common to the line $y = \eta$ and the set E, a set that can be extremely complicated in nature. For this reason, when Harnack [1884–85: II, 286] treated Fubini's Theorem he imposed the customary condition on E that the lines $y = \eta$ meet E in at most two points so that G_η would be an interval. Harnack was unclear as to whether a condition of this sort is essential and later [1886] limited himself to the remark that Fubini's Theorem remains valid for a boundary curve of "sufficiently simple form," that is, lines parallel to the coordinate axes should meet the boundary curve in a finite number of points. The motivation behind this restriction is fairly clear: the inner integrals would then be integrated over a finite sum of intervals, so no new theoretical problems arise concerning their meaning. Domains not satisfying this condition were dismissed by Harnack with the remark that a special investigation would be necessary.

Similarly, Arzelà [1891], although proposing to establish Fubini's Theorem "in its generality" as well as in a manner simpler than

Harnack's, assumed throughout that E is bounded by a simple closed continuous curve that is rectifiable. It may be that he had taken Harnack's warning concerning possible limitations to the validity of this theorem too seriously, for he does in fact assume that the sets G_η are unions of intervals [1891: 136–37]. Actually that is not even true under Arzelà's hypotheses, as the following example illustrates: Let g denote Volterra's function with nonintegrable derivative (3.1, following eqn (6)). Let E be the region bounded by the lines $x=0$, $x=1$, $y=-1$, and $y=g(x)$. Since g' is bounded, $y=g(x)$ is rectifiable and Arzelà's hypotheses are satisfied. But for $\eta=0$, the set G_η consists of the nowhere dense set $\overline{G}$ of Volterra's construction together with an infinite number of points within each of the intervals that form the complement of $\overline{G}$. This example, incidentally, also shows that the sets G_η need not be measurable.

The treatments of double integrals and in particular of Fubini's Theorem show that although the theory of integration had attained precision and generality with regard to the function concept, it continued to lack these qualities when applied to functions of two variables: the domain of integration had not been submitted to the same careful scrutiny accorded the notion of a function. The situation was finally rectified by Jordan [1892] through his development of the concept of a measurable set.[4] The results of Riemann's theory of integration, especially as treated by Darboux, Jordan observed,

> completely clarifies the role which the function plays in the integral.
>
> The influence of the nature of the domain does not appear to have been studied with the same care. All the demonstrations rest upon this double postulatum: that each domain E has a

4. It is difficult to document Jordan's familiarity with the work of his predecessors because he was not in the practice of acknowledging his indebtedness to others, a habit which put him on bad terms with Hermite (see [Hermite 1916: letter of 3 September 1887]). It is clear, however, that Jordan was familiar with Cantor's ideas. According to Hermite [1916: letter of 22 June 1882 (p. 217)], Jordan was proficient in German, so it is likely that he had read the papers of Stolz and Harnack as well. Jules Tannery [1887] had given a favorable review of Peano's *Applicazioni* [1887], singling out the chapter on geometrical magnitudes for special praise. Although Tannery did not enter into details, he did mention that Peano had introduced the notions of internal, external, and limit (i.e., boundary) points of a domain and the distinctions between internal and external length, and between area and volume. Even if Jordan had not read Peano, it is likely that he had read Tannery's suggestive review.

determinate extension; and that, if it is decomposed into several parts E_1, E_2, $\cdots$, the sum of the extensions of these parts is equal to the total extension of E. But these propositions are far from being evident if full generality is allowed to the concept of the domain. [1892: 69–70.]

Jordan thus began, as had Peano, by giving precise, general definitions of the notions of interior, exterior, and boundary points and of the Cantorian notions of "limit point" and "closed set"; but Jordan's definitions were, following Cantor, given in the more general setting of n-dimensional space. The inner and outer content, $c_i(E)$ and $c_e(E)$, of an arbitrary bounded set E were then defined in a manner similar to Peano's, and E is said to be measurable when $c_i(E) = c_e(E)$. Jordan showed that for any set E, if $E = E_1 \cup E_2 \cup \cdots \cup E_n$, where the E_p are mutually disjoint, then

$$\sum_{p=1}^{n} c_i(E_p) \leq c_i(E) \leq c_e(E) \leq \sum_{p=1}^{n} c_e(E_p).$$

Thus, if the E_p are measurable, their sum, E, is likewise measurable, and $c(E) = \sum_{p=1}^{n} c(E_p)$. Jordan was the first to emphasize the importance of the additivity of a measure in the theory of multiple integrals as well as the relation of additivity to the concept of measurability.

Jordan then proceeded to consider the upper and lower Riemann sums, U and L, which correspond to a partition $E = \bigcup_{p=1}^{n} E_p$ of a measurable set E and an arbitrary bounded function f defined on E:

$$U = \sum_{p=1}^{n} M_p c(E_p) \quad \text{and} \quad L = \sum_{p=1}^{n} m_p c(E_p).$$

Using what is essentially Thomae's proof, he showed that U and L approach limits as the dimensions of the E_p approach zero and termed them the "intégrale par excès" and the "intégrale par défaut." Thus, $\int_E f(x,y)dE$ exists if and only if these two integrals coincide.

Although he was primarily interested in the theory of integration in n-dimensional space for $n > 1$, Jordan gave his definitions for arbitrary n. For $n = 1$ they yield a definition of Riemann integrability that employs partitions into arbitrary measurable sets rather than into just intervals. In this form Riemann's definition no longer appears incapable of generalization; or, at least, it is possible to see a relationship between the problem of extending Riemann's definition and the problem of extending the class of measurable sets and the accompanying measure.[5] This relationship is also suggested by Peano's "geometric" interpretation of the integral. Lebesgue's generalization of the integral was based upon the implications of both Peano's and Jordan's formulations of Riemann's definition.

If E is not measurable, Jordan defined $\int_E f(x,y)dE$ as follows: Let $E_1, E_2, \cdots$ be an infinite ascending sequence of measurable sets such that $\lim_{n \to \infty} c(E_n) = c_i(E)$. Then $\int_{E_n} f(x,y)dE_n$ converges, since

$$\left| \int_{E_{n+p}} f - \int_{E_n} f \right| = \left| \int_{E_{n+p} - E_n} f \right| \leq M[c_i(E) - c(E_n)],$$

where M denotes an upper bound for $|f(x,y)|$ on E. Hence, $\int_E f(x,y)dE$ can be defined as $\lim_{n \to \infty} \int_{E_n} f(x,y)dE_n$; similar definitions can be given for upper and lower integrals when f is not integrable on each E_n.

With these definitions behind him Jordan was ready to establish Fubini's Theorem in its full generality. Suppose for simplicity that E is a bounded, measurable subset of the plane and that $f(x,y)$ is integrable over E. Let F denote the set of numbers y such that (x,y) is in E for some value of x, and let G_η be defined, as before, as the set of points common to the line $y = \eta$ and the set E. As Jordan pointed out, G_η need not be measurable. Nevertheless, in accordance with Jordan's extended definition, $\int_{\underline{G_\eta}} f(x,\eta)dx$ and $\int_{\overline{G_\eta}} f(x,\eta)dx$ always have meaning, and Fubini's Theorem becomes:

$$\int_E f(x,y)dE = \int_F d\eta \left[\int_{\underline{G_\eta}} f(x,\eta)dx \right] = \int_F d\eta \left[\int_{\overline{G_\eta}} f(x,\eta)dx \right].$$

This statement of Fubini's Theorem and the proof Jordan gave con-

5. It should be noted, however, that Jordan defined the lower and upper integrals as limits rather than as a least upper bound and a greatest lower bound, as in the Introduction. It is the latter definition, used by Peano, that provides a fruitful generalization.

tain much in common with Arzelà's and Harnack's treatments, but Jordan was the first to turn their ideas into a lucid, entirely general proof. The really essential hypothesis on E, he showed, is measurability. To illustrate this point, Jordan considered the nonmeasurable set E consisting of those points (x,y) with $0 \leq y \leq 1$ such that either y is rational and $0 \leq x \leq 1$ or y is irrational and $-1 \leq x \leq 0$. The theorem fails to hold even for $f(x,y) = k$ for $k \neq 0$ because $\displaystyle\int_E f(x,y)dE = kc_i(E) = 0$ but, as G_η is an interval of length 1 for all η and $F = [0, 1]$,

$$\int_F d\eta \left[\int_{G_\eta} f(x,\eta)dx \right] = k.^6$$

Lebesgue [1926: lxi] referred to Jordan as a "traditionalistic innovator"—a characterization which is particularly apt in connection with his 1892 paper. The notion of measurability and its relevance to the existence of $\displaystyle\underline{\int}_E f$, $\displaystyle\overline{\int}_E f$, and $\displaystyle\int_E f$ as limits and to Fubini's Theorem had been apprehended with varying degrees of insight prior to Jordan's paper; indeed, Jordan's proofs borrow from those of his predecessors. But by placing the discussion within a completely general set-theoretic framework commensurate with Cantor's work and by introducing the concept of measurability, Jordan gave to the theory a clarity, generality, and suggestiveness that it had never before possessed. It was also through his elegant treatment and application of the measure of sets that these ideas became known and were accepted, for they were incorporated into the second edition of his widely read *Cours d'analyse* (Vol 1, 1893). Through *Cours d'analyse* Jordan's ideas influenced Borel and, particularly, Lebesgue.[7] Lebesgue did not meet Jordan personally until after the completion of his doctoral thesis in which he presented his generalizations of Jordan's theory of measure and of Riemann's theory of the integral [Lebesgue 1926: l].

6. If $\int_E f(x,y)dE$ had been defined as $\lim\limits_{n \to \infty} \int_{E_n} f(x,y)dE_n$, where the E_n are measurable, $E_n \supset E_{n+1} \supset E$ for all n, and $\lim\limits_{n \to \infty} c(E_n) = c_e(E)$, the same difficulty arises, since then $\int_E f(x,y)dE = kc_e(E) = 2k$. The set E, however, is Lebesgue-measurable and has measure one. Thus the Lebesgue integral $\int_E f(x,y)dE$ equals k.

7. In Lebesgue's words: "In a sense Jordan rehabilitated this theory [of sets]; he affirmed that it is a useful branch of mathematics. He did more than affirm it, he proved it through his researches on the measure of areas and sets, on integration . . . which have so well prepared certain works, mine in particular" [1922: 15].

4.2   BOREL'S THEORY OF MEASURE

Jordan's 1892 paper definitely placed Riemann's theory of integration within a measure-theoretic context—the context out of which the modern theory of integration was to evolve. Despite the obvious historical significance of the adoption of Jordan's viewpoint, it must be kept in mind that the modern theory of integration is based upon a different conception of measure. Of course, the modern conception of measure is not without relevance to Riemann's theory, since a bounded function is integrable if and only if its points of discontinuity form a set of zero Lebesgue measure. But this fact had been obscured, or more precisely, had been expressed in terms that avoided such a notion. New ideas on the measure of sets came from an entirely different source: Émile Borel's early researches on complex function theory. Certain developments within that theory suggested to Borel the usefulness of a different sort of measure, one that would assign zero measure to certain dense sets.

In 1851 Riemann raised the question as to whether the class of analytic functions is identical with the class of functions of the complex variable z that are defined by applying to z the basic operations—addition, subtraction, multiplication, and division—a finite or infinite number of times [1851: 39]. The question had been raised rather offhandedly, but Weierstrass awakened interest in it by showing, in one of his few publications [1880], that these two conceptions of a function are not equivalent, since the function $f(z) = \sum_{n=0}^{\infty} 1/(z^n + z^{-n})$ represents two distinct analytic functions for $|z| < 1$ and $|z| > 1$. f is unbounded in any neighborhood of a point z_0 with $|z_0| = 1$, so the analytic function $f(z)$ represents for $|z| < 1$ cannot be analytically continued across the circle $|z| = 1$. On the basis of this and more general results of the same nature, Weierstrass made two points: (a) Many of the most important theorems in the modern theory of functions may not be applied to the expressions that earlier analysts such as Lagrange and Euler termed functions of a complex variable; and (b) There exist analytic functions whose points of nondefinability are not discrete but form lines and two-dimensional regions in the plane. [1880: 729, 741.]

Charles Hermite (1822–1901) was greatly impressed with Weierstrass' remarks, and one of his students, Paul Appell (1855–1930), soon constructed further examples exhibiting the same behavior as Weierstrass' but of the form $\sum A_n/(z-a_n)^{m_n}$. (See [Appell 1882a; also 1882b].) Another of Hermite's former students, Henri Poincaré (1854–1912),

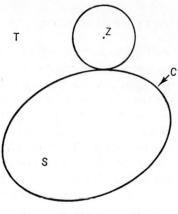

Fig. 3

was encouraged to seek new examples. Poincaré [1883] was able to indicate a general procedure for constructing such examples that becomes especially important in connection with Borel's work. Let C denote a convex contour, a circle for example, and let S and T denote the bounded and unbounded regions into which the plane is divided by C. (See Figure 3). It is assumed that C possesses a tangent and radius of curvature at each point, so for any point z in T there exists a circle with center z which is tangent to C and lies entirely outside of S. Poincaré defined a function f on T by

$$f(z) = \sum_{n=0}^{\infty} \frac{A_n}{z - b_n},$$

where the series $\sum_{n=0}^{\infty} |A_n|$ converges and the points b_n are assumed to form a subset of $C \cup S$ which is everywhere dense in C. The function f is analytic in T and has the property, by virtue of the density of the b_n in C, that its power series expansion about any z in T has for its circle of convergence the tangent circle. Thus, f cannot be continued analytically across C; it possesses the "espace lacunaire" S.

Eduard Goursat (1858–1936) had constructed [1882] this same class of series for the same reason. Later T.-J. Stieltjes (1856–94) also gave a particular example of this type [1887] and Goursat [1887] republished his results. This class of series is particularly intriguing, because when the b_n are on C they also define an analytic function in S. When Poincaré republished his construction [1892], however, he insisted that it could make no sense to regard the *function* $f(z) = \sum\limits_{n=0}^{\infty} A_n/(z-b_n)$—as opposed to the analytical expression to the right of the equality sign—as existing outside of T, presumably because of the impossibility of analytical continuation over C. Further peculiarities of the analytical functions defined by these series were noted by A. Pringsheim (1850–1941) that same year [1892; 1893].

Thus it seemed that these much-discussed series did not fall within the purview of complex function theory. Despite the number and prominence of the mathematicians who espoused this view, Émile Borel (1871–1956) proposed to show in his doctoral thesis of 1894 (for which Appell and Poincaré were *rapporteurs*) that for these series "it is possible, in certain cases, to give a definition of analytic continuation beyond an essentially closed singular line which is neither in contradiction with itself nor with previous notions" [1895: 10]. Borel considered series of the form

$$(3) \qquad\qquad \sum A_n/(z - a_n)^{m_n},$$

where the m_n denote integers bounded above by a fixed integer (to avoid Appell's examples) and the series $\sum |A_n|$ converges. It is assumed that the closure of the set of a_n consists of at most curves and isolated points. It will be instructive to keep Poincaré's paper in mind, as Borel undoubtedly did, and to consider the case in which the points a_n form a dense subset of the curve C in Figure 3. Borel proved that under these conditions the function defined by (3) possesses many of the characteristic properties of analytical functions, including in particular the property expressed by the so-called "Identity Theorem,"[8] which Weierstrass had referred to (in (a) above) as a property that could not be assumed for analytical expressions that do not represent the same analytic function.

Borel's most significant discovery from the standpoint of the development of the theory of measure is (again specialized to the context of Poincaré's paper) that when $\sum |A_n|^{1/2}$ converges, any point in T can be connected to any point in S by a circular arc on which the series (3) with $m_n = 1$ converges absolutely and uniformly, even though the arc meets C. Thus, in a sense, the function defined by (3) can be analytically continued across C to S. Borel's proof of this result suggested the

---

8. If two analytic functions take on the same values on an infinite set of points that has a limit point, they must be identical throughout their common domain of definition.

importance of a definition of measure that employs an infinite number of intervals to cover the set being measured.

The proof proceeds as follows [Borel 1895: 24ff]: Let P denote a point in the region T, and Q a point in S, where T, S, and the curve C are the same as in Poincaré's paper. (Again, the proof is specialized to the context of Poincaré's paper.) Let $\overline{AB}$ denote any segment on the perpendicular bisector of $\overline{PQ}$ and O any point lying on $\overline{AB}$. (See Figure 4.) Each O determines a circle that passes through P and Q. Consider one of the arcs $\widehat{PQ}$. It may happen that $\widehat{PQ}$ meets the curve C in one of the points a_n. Suppose without loss of generality the worse possible case: for every n, the points P, Q, and a_n determine a circle with center O_n lying on $\overline{AB}$. Let L denote the length of $\overline{AB}$. The condition that $\sum |A_n|^{1/2}$ converges implies the existence of a convergent series of positive terms, $\sum u_n$, such that $\sum |A_n| / u_n$ also converges. An integer N can thus be chosen so that $\sum_{n=N+1}^{\infty} u_n < \frac{1}{2}L$. For n greater than N construct the interval I_n on $\overline{AB}$ with center O_n and length $2u_n$. The sum of the lengths is then $2 \sum_{n=N+1}^{\infty} u_n$ and hence is less than L. From this fact Borel was able to deduce the existence of an uncountable number of points of $\overline{AB}$ that lie outside the I_n. (See below.) Thus there exists a point W on

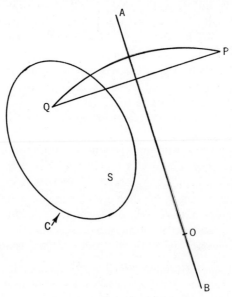

Fig. 4

$\overline{AB}$ that is neither in any I_n nor equal to O_n for $n = 1, 2, \cdots, N$. Consequently, the circle that has W has its center and passes through P and Q contains no a_n. Using that fact, Borel then proved that the series (3) converges absolutely and uniformly on this circle.

Borel's construction of the I_n is reminiscent of Harnack's proof that any countable set can be enclosed in intervals of arbitrarily small total length. Indeed, Borel was fully aware of this property of countable sets and used it to deduce the existence of an uncountable number of points W on $\overline{AB}$ outside the I_n, $n > N$, from the existence of one such point. More precisely, Borel proved that if I_n, $n = 1, 2, 3, \cdots$, is any collection of intervals contained in a fixed interval $[a,b]$ and of total length less than $b - a$, then there exists a point of $[a,b]$ not in any of the I_n.[9] From this proposition he then deduced that an uncountable number of points of $[a,b]$ lie outside the I_n; for if only a countable number of such points existed, they could be enclosed in intervals of total length sufficiently small that their total length plus that of the I_n would still be less than $b - a$. Here we have the above-mentioned property of countable sets being used.

Consider more closely the implications of Borel's result. There exist uncountably many points O on $\overline{AB}$ such that the series (3) converges absolutely and uniformly on the circle determined by P, Q, and O. Since different O's determine different circles, the circles will meet C in different points. Thus, there exist uncountably many points on C for which the series (3) converges absolutely, despite the fact that the singular points a_n are densely distributed on C. This result appears startling in precisely the same manner as does the fact that countable dense sets can be enclosed in intervals of arbitrarily small total length— the fact that is essentially at the heart of Borel's proof and that, according to Borel [1940: 391–94], led him to his conception of a set of measure zero.

9. Borel's proof of this result is also of interest. To establish the existence of one point not in the I_n, he assumed no such point exists. Then it can be further assumed that the I_n, exclusive of their endpoints, contain all of $[a,b]$, since, if not, they can always be slightly expanded in such a manner that their length is still less than $b - a$. Under these conditions, Borel confessed, the conclusion may appear almost self-evident, but because of its importance he proposed to provide a demonstration "based on a theorem interesting by itself . . . : If one has an infinity of subintervals on a line such that every point of the line is interior to at least one of them, a finite number of intervals chosen from among the given intervals can be effectively determined having the same property" [1895: 51]. Here we find the first statement of the so-called Heine-Borel Theorem.

Four years later in the monograph *Leçons sur la théorie des fonctions* [1898], Borel was able to obtain further results by developing the measure-theoretic implications contained in his thesis. In his thesis he had enclosed the points O_n in intervals; in the monograph he dealt directly with the points a_n and studied more closely the nature of the points of convergence of the series (3). Since

$$\left| A_n/(z - a_n)^{m_n} \right| = \left| A_n \right| / [(x_1 - a_1^{(n)})^2 + (x_2 - a_2^{(n)})^2]^{m_n/2},$$

where $z = x_1 + i x_2$ and $a_n = a_1^{(n)} + i a_2^{(n)}$, the problem is a special case of the problem of determining the nature of the points of convergence of the real-valued series with nonnegative terms

$$A_n/[(x_1 - a_1^{(n)})^2 + \cdots + (x_m - a_m^{(n)})^2]^{m_n/2}.$$

For the sake of simplicity, we shall follow Borel by considering only the simplest case in detail: $m_n = 1$, $m = 1$, and, for the series

$$(4) \qquad\qquad \sum A_n/\left| x - a_n \right|,$$

the real numbers a_n lie in $[0,1]$.

As in his thesis, Borel began by assuming the convergence of the series $\sum A_n^{1/2}$, so there exists a series with positive terms u_n such that the series $\sum A_n/u_n$ also converges. If $v_n = A_n/u_n$ and x is a point in $[0,1]$ such that $\left| x - a_n \right| > v_n$, then

$$(5) \qquad\qquad \frac{A_n}{\left| x - a_n \right|} < u_n,$$

so x is a point of convergence of the series (4) if x is not in any of the intervals $[a_n - v_n, a_n + v_n]$. The total length of these intervals is $2 \sum v_n = 2v$, and, by virtue of (5), (4) converges uniformly on the subset of $[0,1]$ that is complementary to the union of these intervals. Next Borel considered what happens when the series $\sum u_n$ is replaced by the series with terms $u_n' = 2k u_n$. Then

$$\sum v_n' = \sum \frac{A_n}{u_n'} = \frac{1}{2k} \sum v_n,$$

so the series $\sum v_n'$ converges. It follows as before that if x is not in any of the intervals $[a_n - v_n', a_n + v_n']$, the series (4) will converge, and again the convergence is uniform. The sum of the lengths of these intervals, $2 \sum v_n' = v/k$, can consequently be made arbitrarily small by choosing k sufficiently large. In other words, (4) converges uniformly on sets with Lebesgue measure arbitrarily close to 1 and diverges on a set of zero measure.

These results and their application to the complex series (3) take up the second half of Borel's monograph. Borel was convinced that the set-

theoretic and measure-theoretic ideas involved were sufficiently important to warrant a careful treatment, which he undertook in the first
half. Before we discuss its contents, however, a few prefatory remarks
are in order concerning the work of Jules Drach, a friend and fellow student of Borel's at the École Normale Supérieure whose ideas had an important influence upon both Borel and Lebesgue.

In 1895 Borel and Drach published a book based on Jules Tannery's
lectures at the École Normale on the theory of numbers and higher
algebra. Drach treated the higher algebra in a manner characteristically
his own. "The mode of exposition," wrote Tannery in the preface, "to
which he [Drach] has been led by the desire to reduce the construction
of Arithmetic and Algebra to what is indispensable consists essentially
in regarding the positive or negative integers and rational numbers as
signs or *symbols*, entirely defined by a small number of properties posited
a priori relative to two of their modes of composition" [1895: iv]. In
other words, Drach's method amounted to taking an abstract or postulational approach to a particular field of mathematics. Drach applied
this approach in his doctoral thesis [1898; cf. 1895] to the study of differential equations, the integral solutions being characterized and classified on the basis of essential properties which they must possess.
Drach's approach was not without precedent, particularly outside of
France, but, historically, his work was instrumental in initiating the
axiomatic approach to the theory of measure and integration; this
approach, in turn, focused attention upon what properties or charac-
teristics a viable definition of measure and the integral should possess.

The consequences of taking the axiomatic approach are particularly
evident in Borel's concept of measurability. For simplicity it is assumed
that all sets are subsets of [0,1]. Measurable sets and their measure are
then defined in the following manner:

> When a set is formed of all the points comprised in a denumer-
> able infinity of intervals which do not overlap and have total
> length s, we shall say that the set *has measure s*. When two sets
> do not have common points and their measures are s and s', the
> set obtained by uniting them, that is to say their sum, has
> measure $s+s'$.
>
> More generally, if one has a denumerable infinity of sets which
> pairwise have no common point and having measures $s_1, s_2, \cdots,$
> $s_n, \cdots$, their sum . . . has measure
>
> $$s_1 + s_2 + \cdots + s_n + \cdots.$$

All that is a consequence of the definition of measure. Now
here are some new definitions: If a set E has measure s and con-

tains all the points of a set E' of which the measure is s', *the set*
$E-E'$, *formed of all the points of E which do not belong to E', will
be said to have measure $s-s'$* . . . .

  *The sets for which measure can be defined by virtue of the preced-
ing definitions will be termed measurable sets by us* . . . . [1898:
46–48.]

"The procedure that we have employed," Borel added concerning the
nature of his definition,

> actually amounts to this: we have recognized that a definition
> of measure could only be useful if it had certain fundamental
> properties; we have posited these properties *a priori* and we
> have used them to define the class of sets which we regard as
> measurable. This manner of proceeding presents extensive
> analogies with the methods introduced by M. J. Drach, in
> Algebra and in the theory of differential equations . . . . In any
> case, it proceeds from the same fundamental idea: define the new
> elements which are introduced with the aid of their *essential*
> properties, that is to say those which are strictly indispensable
> for the reasoning that is to follow. [1898: 48 n1.]

Indeed, we have already seen the manner in which a countable infinity
of covering intervals is "strictly indispensable" for Borel's analysis of
the behavior of the series (4). When the a_n are dense in [0,1], the set of
points of convergence and the set of points of divergence of (4) both
have outer content 1 and inner content 0, and the distinct nature of
the two sets is not apparent. Undoubtedly Borel had in mind the theory
of content when he remarked that another definition of measure is
possible, "but such a definition would be useless to us; it could even be
a hindrance if it did not allow to measure the fundamental properties
that we attributed to it in the definition we gave" [1898: 48].

  The fundamental property behind the application to (4), as Borel
noted, is that every countable set has zero measure—the property that
had originally suggested the new definition of measure. Borel did not
bother to clarify the general nature and definition of his measurable
sets, although he did point out that every closed set is measurable,
since, as the results of Cantor and Bendixson showed, it is the complement of a countable number of nonoverlapping intervals (see 3.3). In
the application of his theory of measure, however, Borel was con-
fronted with more complicated sets, such as the set D of noncon-
vergence points of the series (4). Actually D is measurable in Borel's

sense, but he was unsure on this point [1898: 67n]. It was clearly because of D that he added the following remark after presenting his definitions of measure and measurability:

> However, if a set E contains all the elements of a set E_1, of measure α, we can say that the measure of E is greater than α, without worrying whether E is measurable or not. Conversely, if E_1 contains all the elements of E, we will say that the measure of E is less than α. The words *greater* and *less*, moreover, do not exclude equality. It is easy to see that the essential properties extend, with suitable modifications, to these new definitions: in a sense, a calculus of equalities is replaced by a calculus of inequalities which sometimes renders the same services. [1898: 48–49.]

With this convention, therefore, D has measure less than or equal to zero, "but measure is never negative" [1898: 67n]. Hidden away in Borel's convention—introduced more or less as a necessary evil because of D—is the notion of a Lebesgue-measurable set. More precisely, if E is a set such that $E_1 \subset E \subset E_2$, where E_1 and E_2 are measurable in Borel's sense and have the same measure α, then Borel's convention about inequalities requires that the measure α also be assigned to E; the class of sets which are assigned a measure in this manner coincides with the class of Lebesgue-measurable sets—a fact pointed out by Lebesgue (see 5.1).

Nowhere in the *Leçons* did Borel suggest a connection between his concept of measure and the theory of integration. The reason for this is reflected in his comment: "It will be fruitful to compare the definitions that we have given with the more general definitions that M. Jordan gives in his *Cours d'analyse*. The problem we investigate here is, moreover, totally different from the one resolved by M. Jordan . . ." [1898: 46n]. Borel thus considered Jordan's definitions more general, which is indeed the case in the sense that there are many Jordan-measurable sets which are not Borel-measurable.[10] Borel's definition (without his convention) is less general because it takes into consideration the structure of the sets being measured, that is, their composition out of intervals, a quality which he found philosophically attractive [cf.

---

10. This can be seen in the following manner. If E is a perfect, nowhere dense set of zero content, then every subset also has zero content and is therefore measurable in Jordan's sense. E has the cardinality c of the continuum. Thus the class of Jordan-measurable sets has cardinality 2^c. On the other hand, the class of Borel sets has cardinality c.

Borel 1912: 28–29]. Although some non-Jordan-measurable sets do become Borel-measurable, Borel's theory of measure was not intended as a generalization of the theory of content. Being regarded in this light, Borel's theory would obviously not seem appropriate for generalizing the integral concept. Furthermore, as Borel himself emphasized in the above quotation, the application that he had in mind—the behavior of the series (4)—was far removed from the problems of the theory of integration.

4.3   SCHOENFLIES' REPORT ON THE THEORY OF SETS

Towards the end of the nineteenth century, the German mathematician Arthur Schoenflies (1853–1928) was commissioned by the Deutsche Mathematiker-Vereinigung to prepare a report on curves and point sets. A preliminary report was presented in September of 1898 and eventually grew into a two-volume work on the theory of sets and transfinite numbers. Volume 1 [Schoenflies 1900] represents the first treatise on set theory. Over half of this work is devoted to applications to the theory of functions of a real variable, including the theory of integration. Although it represents the work of one man, it probably accurately reflects the attitudes prevalent at the turn of the century—when Lebesgue's own research was getting under way—towards Borel's theory of measure and towards the "problems" that had arisen within Riemann's theory of integration.

When he came to compose the section of the report dealing with the measure of point sets, Schoenflies was faced with the fact that three theories were then in existence: the theory of outer content of Stolz, Cantor, and Harnack; the theory of content of Peano and Jordan; and Borel's theory of measure. Thus Schoenflies began by remarking that a definition of measure possesses, as does every mathematical definition, a certain subjective character and must be judged by the degree to which its consequences are suited to the objectives behind its introduction. We shall see that Schoenflies questioned (in part by implication) the tenability of Borel's definition in this connection. That is, Borel's definition appeared to Schoenflies as being neither of any use in most contemporary applications of the theory of content nor necessary for, or even appropriate to, Borel's own application to the study of the series $\sum A_n/|x-a_n|$.

We shall begin, however, by considering another objection of Schoenflies. A characteristic property of outer content, Schoenflies observed, is that the limit points of a set are taken into consideration in determining the extension of the set. This property, of course, is true in Peano's and Jordan's theory, which simply represents a refinement of the theory of outer content. "E. Borel has recently taken an essentially different point of view. To a set P he does not add its limit points; besides this, he also disregards the requirement of a finite number of regions which contain all points of a set P . . ." [1900: 88]. These radical departures from the theory of content did not appeal to Schoenflies, especially in the postulational form that Borel presented his definition. Borel's stipulation that his measure be countably additive, Schoenflies remarked, "only has the character of a postulate since the question of whether a property is extendable from finite to infinite sums cannot be settled by positing it but rather requires investigation" [1900: 93]. Here Schoenflies objected to the form of Borel's definition with its concomitant indefiniteness. It still has to be determined, he seems to be saying, whether a definition of measure similar in form to Jordan's can be given which will satisfy the conditions set down by Borel. However, Schoenflies did not consider countable additivity to be the property *sine qua non* for a measure that it has since become.

In connection with Schoenflies' implication that Borel's definition was of no use in most contemporary applications of the theory of content, it must be kept in mind that the theory of content originally evolved from the theory of Riemann integration and that that remained its principal field of application. In particular, Riemann's integrability condition had become: f is integrable if and only if S_k, the set of points at which the oscillation of f is greater than or equal to k, forms "an extensionless set" for every positive k. After presenting a proof of this proposition, Schoenflies commented (with Borel's definition obviously in mind) that "here only that concept of content comes into question that operates with a finite number of subintervals and adds the limit points to a set" [1900: 181n]. This appears to be a rather strange remark, since Schoenflies [1900: 127–28, 180–81] realized that the sets S_k are closed and that for closed sets Borel measure agrees with outer content. So accustomed had mathematicians become to the form of Riemann's condition that they, or at least Schoenflies, failed to see that Borel's definition affords a much neater characterization of Riemann integrability. Once it is realized that Borel measure and outer content agree on the S_k, it is not difficult to see that f is integrable if and only if the S_k have Borel measure zero and that this is so if and only if the set of points of discontinuity, $D = \bigcup_{n=1}^{\infty} S_{1/n}$, has Borel measure zero. Schoenflies' blindness to this result was probably due in part to the fact that he peremptorily dismissed Borel's measure.

One of Schoenflies' reasons for not considering Borel's definition as a serious candidate was probably that, by ignoring limit points, it made everywhere dense sets extensionless. The lack of appeal of this consequence is also evident in connection with Schoenflies' implication that Borel's definition was not appropriate to Borel's own studies. In his study of the series $\sum A_n/|x-a_n|$, Borel had proved that the measure of D, the set of points of nonconvergence, is zero. When Schoenflies [1900: 243 ff] treated Borel's researches, however, he made no mention of Borel's measure. In fact Schoenflies had earlier made the convention that "content" was to be used to refer to outer content since "for applications it is always only a question of this outer content . . ." [1900: 94]. Thus he notes that $c_e(C) = 1$, where C denotes the points of convergence. It has already been noted that since $c_e(D) = 1$, nothing much is revealed about the distinct natures of C and D. On the other hand, D is characterized as being contained in a dense, uncountable set that is of the second category[11] rather than being in a set of Borel measure zero. That is, (in the notation of 4.2) let U_k denote the union of the intervals $[a_n - v_n', a_n + v_n']$ outside of which the series converges uniformly, where $v_n' = v_n/k$. Then, since the points a_n are assumed to be dense in [0,1], V_k, the complement of U_k, is nowhere dense; consequently, $M = \bigcup_{k=1}^{\infty} V_k$ is of the first category. Therefore, D is contained in the set $\bigcap_{k=1}^{\infty} U_k = [0,1] - M$, a set of the second category. Thus, although $\bigcap_{k=1}^{\infty} U_k$ is a "Borel set" of zero measure, Schoenflies stressed its "bigness" by showing that it is a second category set. He seems to have tacitly rejected the idea that such a set should be regarded as extensionless.

"The fundamental theorem of higher analysis," remarked Schoenflies by way of introducing the section on the Fundamental Theorem, "states that in general differentiation and integration are inverse operations" [1900: 206]. The qualification "in general" had to be added since (in view of the developments discussed in Chapter 3), if arbitrary functions are considered, the Fundamental Theorem fails to hold in a number of respects. In particular, Volterra had indicated a method for constructing a class of functions with bounded, nonintegrable derivatives (Ch. 3). By the time Schoenflies composed his report, another

---

11.   A set is of the first category if it can be represented as a countable union of nowhere dense sets. A second category set is one which is not of the first category. These notions were introduced by Baire in 1898.

class of functions possessing the same property had been discovered and were much discussed.

Volterra's discovery had been motivated by some conjectures of Dini, including one to the effect that there exist functions that oscillate infinitely often in every interval and yet possess a bounded derivative (2.3). By means of geometrical construction H. A. Köpcke [1887] succeeded in confirming the truth of Dini's conjecture.[12] Köpcke was not particularly interested in the fact that f' in his example is not integrable, but in the fact that his example extended the class of nonintuitable continuous functions, which had formerly been regarded as occupied solely by nondifferentiable functions. A simplified construction of Köpcke's example was given later by I. Pereno [1897].

The Swedish mathematician Torsten Brodén pointed out [1896] that Cantor's principle of condensation of singularities (3.4) affords a simpler example of the type of function discovered by Köpcke. Brodén considered the function

$$f(x) = \sum_{n=1}^{\infty} c_n g(x - w_n),$$

where: the w_n are dense in $[-1,1]$; the c_n are positive; g is a strictly increasing continuous function; $g'(x)$ is continuous except at 0; and $g'(0) = +\infty$. Actually Brodén was thinking primarily of the function $g(x) = x^{1/3}$, and his conclusions do not go through for the more general case. (Both Weierstrass and Cantor had specifically considered $g(x) = x^{1/3}$.) Brodén observed that since f is a strictly increasing function, it has an inverse $h(x) = f^{-1}(x)$ defined on $[a, b] = f([-1,1])$. Furthermore, he claimed, since $f'(x)$ is determinate for all x (although, of course, infinite for some x), so is $h'(y)$; in fact, $h'(y) = 0$ for the dense set of y of the form $y = f(w_n)$. He failed to realize, however, that the determinateness of $f'(x)$ for *all* x does not follow immediately from Cantor's principle. But his conclusions are valid for the example he had in mind,[13] although they require some proving. Hence with $g(x) = x^{1/3}$, h' is not integrable although it is clearly bounded on $[a,b]$. It is interesting to observe that Brodén did not mention this property of h. For both Brodén

12. Some of Köpcke's reasoning was incorrect and was later rectified [1889; 1890].
13. This was first established (without knowledge of Brodén's paper) by D. Pompeiu [1906–07], who showed that $f'(x) = \Sigma c_n g'(x - w_n)$ for all x; that is, $f'(x) = +\infty$ if and only if the right-hand side diverges, and $f'(x)$ is finite and equal to the right-hand side if and only if it converges. He also pointed out using Borel's results that if $\Sigma c_n^{1/2}$ converges, then $f'(x)$ is finite except on a set of zero measure.

and Köpcke, the significance of these functions was that they showed an everywhere oscillating function could be everywhere differentiable.

Later Brodén devoted a lengthy paper [1900] to constructing another class of these functions by a method simpler than Köpcke's. Schoenflies himself contributed to the literature on these functions by determining [1901] very complicated conditions on the numbers $f(e_i)$, where the e_i are dense in a fixed interval $[a,b]$ so that the $f(e_i)$ represent the extrema of a function f that is everywhere differentiable on $[a,b]$. Thus by 1900, as Schoenflies remarked in his report, vast classes of functions with bounded but nonintegrable derivatives were known to exist and had been the subject of several papers.

Schoenflies also noted in passing that the upper and lower integrals of a bounded function are analogous to the outer and inner contents of a set; and in 1901 Charles de la Vallée-Poussin expressed this relation more suggestively by defining $c_i(E)$ and $c_e(E)$ to be, respectively, $\underline{\int} \chi_E(x)dx$ and $\overline{\int} \chi_E(x)dx$ [1903: 221]. The connection between measure and integration could not have been stated more clearly, but no one until Lebesgue had the creative imagination necessary to perceive, in this relationship and Borel's conception of measure, the makings of a powerful and useful generalization of the Riemann integral.

4.4   RESEARCHES ON THE TERM-BY-TERM INTEGRATION OF NONUNIFORMLY CONVERGING SERIES

Lebesgue's work is characterized, in part, by its effective use of term-by-term integration of sequences and series that may not converge uniformly. In the 1870's, when the problem of integrating series term by term first came into prominence, the importance of uniform convergence had been overemphasized (see 2.1). By the end of the century, however, attention was focused on the possibility of term-by-term integration of nonuniformly converging series. It was discovered that the validity of term-by-term integration could be proved with the condition of uniform convergence replaced by that of uniform boundedness. But it was then also necessary to impose an additional assumption (continuity or integrability) on the function to which the series converges. The complexity of the proofs—caused by these additional assumptions—stands in sharp contrast with the simplicity and greater generality later obtained by Lebesgue, who was able to build upon the work of his predecessors.

The question of term-by-term integration is, as we have seen, of particular theoretical importance in the theory of trigonometric series. It was in the theory of Fourier series that the nonessential nature of uniform convergence for term-by-term integration became especially evident. Ulisse Dini obtained the first result of this sort in a little-known paper [1874]. A more general result was later obtained independently by Paul du Bois-Reymond [1883b], who showed that any Fourier series corresponding to a function that is Riemann-integrable or that possesses an absolutely converging improper integral (as defined by du Bois-Reymond; see 3.3) can be integrated term by term. Thus, term-by-term integration appeared to be generally valid for Fourier series regardless of the nature of the convergence.

In a subsequent paper du Bois-Reymond [1886] turned to the problem of term-by-term integration of more general infinite series which "ever since Hr. Weierstrass pointed it out, has not been satisfactorily resolved since its solution relative to trigonometric series does not appear to afford an extension to the general case" [1886: 360]. To make some headway in this case, he introduced some distinctions concerning the manner in which a series of continuous functions $\sum_{n=1}^{\infty} u_k(x)$ converges. Let $S_n(x) = \sum_{k=1}^{n} u_k(x)$ and $R_n(x) = \sum_{k=n+1}^{\infty} u_k(x)$. A point x_0 is said to be a continuity point of convergence if $R_n(x) \to 0$ as both $x \to x_0$ and $n \to \infty$, that is, if for every positive ϵ there exists an integer $N = N(\epsilon)$ and a positive $\delta = \delta(\epsilon)$ such that $n \geq N$ and $|x - x_0| < \delta$ implies that $|R_n(x)| < \epsilon$.

Instead of working with $R_n(x)$, du Bois-Reymond chose to consider the analogous case of a function $\phi(x,\epsilon)$ where x is in $[a,b]$, ϵ is positive, ϕ is continuous in x for each value of ϵ, and $\lim_{\epsilon \searrow 0} \phi(x,\epsilon) = 0$ for each x. The problem of term-by-term integration then takes the form: When is it true that, for every subinterval $[x_0,x_1]$ of $[a,b]$,

$$(6) \qquad \lim_{\epsilon \searrow 0} \int_{x_0}^{x_1} \phi(x,\epsilon)dx = 0?$$

He distinguished two cases: (a) There exists a number B such that $|\phi(x,\epsilon)| \leq B$ for all x and all ϵ; (b) No such B exists. He illustrated by examples that in the latter case (6) may or may not be true.

The first case focuses attention on the property of uniform boundedness, which was subsequently discovered to be especially important for the problem of term-by-term integration. This case had already been discussed briefly in 1878 by Leopold Kronecker (1823–91), who main-

tained that a necessary and sufficient condition for (6) is: Correspond-
ing to every pair of positive numbers τ and σ, a positive number ϵ_0 can
be determined so that when $0 < \epsilon \leq \epsilon_0$, the set $E(\epsilon,\sigma)$ (consisting of all
x for which $|\phi(x,\epsilon)| > \sigma$) can be enclosed in a finite number of intervals
of total length less than τ [1878: 54]. In other words, for every σ,

$$\lim_{\epsilon \searrow 0} c_e[E(\epsilon,\sigma)] = 0.$$

Expressed in this form, Kronecker's condition is seen to be the requirement that $\phi(x,\epsilon)$ converges to 0 in measure with respect to outer content. This appears to be the first time this requirement was introduced. The sufficiency of this condition is clear, because if U denotes
the union of the intervals enclosing $E(\epsilon,\sigma)$ then

$$\int_{x_0}^{x_1} |\phi(x,\epsilon)| \, dx = \int_U |\phi(x,\epsilon)| \, dx + \int_{U^c} |\phi(x,\epsilon)| \, dx$$

$$< B\tau + \sigma(b - a),$$

where $U^c = [x_0,x_1] - U$. Its necessity is not quite as apparent, but
Kronecker provided no demonstration. Actually the proposition is mis-
leading, since it implies that an additional hypothesis is necessary for
the validity of (6) in case (a). The results of Arzelà and Osgood were to
show that the continuity or integrability of ϕ is sufficient to guarantee
(6) and, hence, Kronecker's condition is actually superfluous.

Du Bois-Reymond, however, failed to realize that. He was convinced
that "with great likelihood" uniform boundedness does not by itself
guarantee (6) [1887: 358]. Apparently he was led to this opinion because of an example he had constructed of a function $\phi(x,\epsilon)$ such that
ϕ converges to 0 with ϵ for every fixed value of x in $[x_0,x_1]$ but corresponding to any such x are sequences $x_n \to x$ and $\epsilon_n \to 0$ such that
$\lim_{n \to \infty} |\phi(x_n,\epsilon_n)| = \infty$. In fact, this example represents the highlight of

du Bois-Reymond's 1886 paper, although it was later shown to be
invalid. Judging by some of his remarks [1886: 366], du Bois-Reymond
evidently thought that he could obtain a uniformly bounded example
that violates (6) by multiplying his original example ϕ by a "discontinuity factor." The seeming hopelessness of the situation is reflected
in a letter du Bois-Reymond wrote to Kronecker on April 22, 1886,
enclosing a copy of his 1886 paper. After noting that, owing to limitations of space and time, he had only stated Kronecker's condition without discussing it more fully, he promised to make up for that in a subsequent paper. "Then I shall turn my back on this type of mathematics. The results are not worth the effort, and, besides, they do not

stimulate further research; on the contrary, their main effect is to stop research in a certain direction . . . " [Kronecker 1889: 353].

It is, consequently, not entirely surprising to find that little interest was shown in the problem of term-by-term integration of nonuniformly converging series and that the errors in du Bois-Reymond's paper went unnoticed. In 1897, however, W. F. Osgood (1864–1943), professor of mathematics at Harvard University, reawakened an interest in the subject with a profound analysis of the behavior of nonuniformly converging series [1897; cf. 1896]. Osgood had studied du Bois-Reymond's paper carefully. DuBois-Reymond's influence is reflected in the conceptual tools used by Osgood and, possibly, in the fact that Osgood also limited his study to series $f(x) = \sum_{n=1}^{\infty} u_n(x)$ for which the u_n and f are continuous on an interval $[a,b]$. In any case, the continuity assumptions played a vital role in his reasoning.

Osgood defined a point x_0 in $[a,b]$ to be a "γ-point" with respect to the positive number A if for every integer $m > 0$ and every $\delta > 0$ there is always an x in $(x_0 - \delta, x_0 + \delta)$ and an $n > m$ such that $|R_n(x)| > A$. It should be noted that x_0 is a γ-point if and only if x_0 is a discontinuity point of convergence in du Bois-Reymond's terminology. Du Bois-Reymond had believed that the γ-points corresponding to a fixed A could fill up an interval, but Osgood showed, on the contrary, that G_A, the set of γ-points corresponding to A, is always closed and nowhere dense. (The proof of this result depends on the continuity of the functions $R_n(x)$ and hence on the continuity of the u_n and f.) If x_0 is such that $R_n(x)$ becomes unbounded in arbitrarily small neighborhoods of x_0, then x_0 is termed an X-point. These points had been introduced in du Bois-Reymond's example in which every point was supposedly an X-point. But the set of X-points, since it clearly forms a subset of the sets G_A of γ-points, must in fact be nowhere dense in $[a,b]$.

Osgood's main result is that if there are no X-points—that is, if $R_n(x)$ is uniformly bounded—then $\lim_{n \to \infty} \int_{x_0}^{x_1} R_n(x)dx = 0$ for every $[x_0,x_1]$ in $[a,b]$ and term-by-term integration is always permissible. The method of proof is *reductio ad absurdum*: Suppose there is a subsequence n_i such that $\left| \int_{x_0}^{x_1} R_{n_i}(x)dx \right| \geq \epsilon$ for $i = 1, 2, 3, \cdots$; let A be such that $0 < A < \epsilon/(x_1 - x_0)$. Then $\left| \int_{x_0}^{x_1} R_{n_i}(x)dx \right| > A(x_1 - x_0)$ and, without loss of generality, it can be assumed that

(7)
$$\int_{x_0}^{x_1} R_{n_i}(x)dx > A(x_1 - x_0)$$

for $i = 1, 2, 3, \cdots$. In geometrical terms (7) implies that the area of the region C_n bounded by $y = R_n(x)$ and the lines $y = A$, $y = B$ (the shaded region in Figure 5) cannot approach zero as $n_i \to \infty$. (If the area of C_{n_i} approached zero, then $\int_{x_0}^{x_1} R_{n_i}(x)dx$ would necessarily approach a quantity less than or equal to $A(x_1 - x_0)$, the area of the rectangle determined by $x = x_0$, $x = x_1$, $y = 0$, and $y = A$.)

The remainder of Osgood's proof consisted in showing that the area of C_n actually converges to 0 as $n \to \infty$. In this connection the set G_A of γ-points plays an important role. If x is in $[x_0, x_1] - G_A$, there exist an open interval I_x that contains x and an integer N_x such that $|R_n(x')| \le A$ for all x' in I_x and $n \ge N_x$. To make the gist of Osgood's proof more apparent, assume for the moment that a finite number of I_x cover $[x_0, x_1] - G_A$, so for all x in this set,

(8)
$$|R_n(x)| \le A \quad \text{if } n \ge N,$$

for a sufficiently large integer N. That implies that the curves $y = R_{N+p}(x)$, for $p = 0, 1, 2, \cdots$, rise above the line $y = A$ and contribute to C_{N+p} only when x is in G_A. For each such x, let $j = j(x)$ denote the smallest integer j such that $|R_n(x)| \le A$ for all $n \ge j$. To study the behavior of $y = R_{N+p}(x)$ on G_A, Osgood introduced the set $G_i (i = 1, 2, 3, \cdots)$, which consists of all x in G_A for which $j(x) \le i$. The significance of the G_i is that $|R_{i+p}(x)| \le A$, $p = 0, 1, 2, \cdots$, for all x in G_i. Thus, the

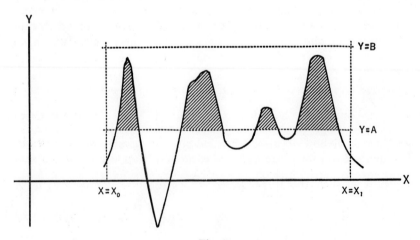

Fig. 5

curves $y = R_{N+p}(x)$ lie above $y = A$ only when x belongs to $G_A - G_N$.

The G_i, like G_A, are closed and nowhere dense; furthermore, it is easily seen that G_i is contained in G_{i+1} and that the union of all the G_i is precisely G_A. Osgood could thus apply the following crucial proposition:

> LEMMA. *If G is a closed, nowhere dense subset of* [a,b] *and* G_i *is a sequence of closed sets such that* $G_i \subset G_{i+1}$ *and* $G = \bigcup\limits_{i=1}^{\infty} G_i$, *then* $\lim\limits_{i \to \infty} c_e(G_i) = c_e(G)$.

This is the first proposition of its kind. Lebesgue used an analogous— but much more general—lemma in a similar but less complicated manner to prove his theorem on term-by-term integration (5.1, Theorem 5.3) and was undoubtedly influenced by Osgood's paper. Osgood's lemma also helped motivate W. H. Young's generalization of the theory of content (5.2). Osgood stressed that this result is "far from being self-evident" [1897: 178], since it is not always valid unless G is itself closed. If each G_i is taken to be a finite set of rational numbers such that $G = \bigcup\limits_{i=1}^{\infty} G_i$ is the set of all rational numbers in [0,1], then $c_e(G_i) = 0$ for all i but $c_e(G) = 1$.

From a heuristic standpoint, one can see that C_{N+p} is entirely contained in the part of the "rectangle" with base $G_A - G_{N+p}$ and height B that is above the line $y = A$; so

$$(9) \quad area(C_{N+p}) \leq (B-A)c_e(G_A - G_{N+p}) = (B-A)[c_e(G_A) - c_e(G_{N+p})];$$

then applying the lemma, $\lim\limits_{p \to \infty} area\ (C_{N+p}) = 0$. This line of reasoning represents the essence of Osgood's proof. To make the proof rigorous, Osgood had to replace G_A and G_{N+p} by two finite unions of intervals whose contents approximate those of G_A and G_{N+p} and for which the reasoning leading to (8) and (9) is valid. (Osgood used the fact that G_A is closed and nowhere dense, so its complement is a countable union of open intervals with endpoints in G_A.) These necessary modifications make Osgood's actual proof long and difficult. Lebesgue was able to give a simple, three-line proof of a much more general result by making use of the fact that the sort of lemma used by Osgood can be established under much more general conditions when Lebesgue measure is substituted for content. Nevertheless, Osgood's paper clearly prepared the way for Lebesgue, and Osgood must be credited with the brilliance that was obviously required to build a proof upon du Bois-Reymond's dis-

tinction between continuity and discontinuity points of convergence.
In a series of notes included in the 1885 *Rendiconto* of the Accademia
dei Lincei of Rome [1885a, 1885b, 1885c], Cesare Arzelà had actually
obtained more general results than Osgood, but they were not widely
known until after their republication [Arzelà 1899–1900]. When Heine
called attention to the importance of uniform convergence in 1870
(2.1), he raised the question whether uniform convergence of an infinite
series of continuous functions is necessary as well as sufficient for the
continuity of the limit function. Shortly thereafter, examples were
published showing that this is not the case. Then in 1883–84 Arzelà
was able to characterize the mode of convergence that preserves continuity in the limit. In 1885 he turned to the analogous problem: If
$f_s(x)$ denotes a sequence of integrable functions on $[a,b]$ such that
$f(x) = \lim_{s \to \infty} f_s(x)$ exists, what are the necessary and sufficient conditions
under which f is also integrable? Arzelà was also familiar with
Harnack's work on trigonometric series and may have been motivated
by Harnack's admission that his proof of the term-by-term integrability of these series fails because the limit function need not be integrable (3.1).

Arzelà's solution is based upon the following result:

> LEMMA. *Let d_n denote a sequence of positive numbers. Suppose
> that corresponding to every n there is a finite number of intervals
> in $[a,b]$ whose union, U_n, has length d_n. Then if $d_n \geq d$ for all n
> where d is positive, there exists a point x_0 in $[a,b]$ with the property
> that x_0 belongs to U_n for infinitely many values of n.*

Since we shall be comparing Arzelà's proof with Lebesgue's,
it should be noted that the proof of this lemma covers four oversized
pages.[14]

This lemma was used to investigate the consequences of the assumption that the limit function f is integrable. Arzelà considered $E(s,\sigma)$,
the set of points x for which $|f_s(x) - f(x)| \geq \sigma$, where σ denotes a fixed
positive number. His first discovery, and the most important from our
viewpoint, is:

(10) $$\lim_{s \to \infty} c_e[E(s,\sigma)] = 0.$$

(This is, of course, Kronecker's condition.) If, for each σ and every s,

---

14. Arzelà's Lemma, which asserts that $\limsup_{n \to \infty} U_n \neq \phi$, follows immediately
from the inequalities $d \leq \liminf_{n \to \infty} m(U_n) \leq \limsup_{n \to \infty} m(U_n) \leq m(\limsup_{n \to \infty} U_n)$,
where m denotes Lebesgue measure. Cf. note 4 to Chapter 6.

the set $E(s,\sigma)$ is the union of a finite number of intervals, the lemma immediately implies (10) since the existence of a point x_0 in $E(s,\sigma)$ for infinitely many values of s is ruled out by the fact that $f(x_0) = \lim_{s \to \infty} f_s(x_0)$.

Because the sets $E(s,\sigma)$ need not possess such a simple structure, Arzelà had to make use of the integrability of the f_s and of f. Essentially, the idea is to show that $E(s,\sigma)$ can be considered as the union of two nonoverlapping sets such that one set satisfies the conditions of the lemma and the other set is of arbitrarily small content. The resulting proof is based upon what amount to compactness arguments—as opposed to purely measure-theoretic arguments—to obtain the "finiteness" commensurate with the definition of outer content; in particular, for the reasoning to be rigorous, it becomes necessary to replace $E(s,\sigma)$ by a finite union of intervals which approximate $E(s,\sigma)$ in content. The result is, as with Osgood, an extremely long and complicated proof.

As observed in connection with Kronecker's paper and as pointed out by Arzelà, (10) immediately implies that term-by-term integration is valid, viz.,

$$\lim_{s \to \infty} \int_{x_0}^{x_1} f_s = \int_{x_0}^{x_1} f$$

for all $[x_0, x_1]$ contained in $[a,b]$, provided that the f_s are uniformly bounded. Using (10), Arzelà also determined the necessary and sufficient condition on the mode of convergence that insures the integrability of f. That condition is complicated, but Arzelà's results also show that the limit function f will be Riemann-integrable if and only if (10) is valid.

Although Arzelà did not offer any examples, it seems that he believed that (10) need not hold, and hence, that f need not be integrable. Indeed, Volterra's example of a function g with bounded nonintegrable derivative g' (3.1) makes that clear: if $\lim_{s \to \infty} h_s = 0$, the sequence of functions

$$f_s(x) = \frac{g(x + h_s) - g(x)}{h_s}$$

converges to $g'(x)$, and the f_s are uniformly bounded because $g'(x)$ is bounded.[15] A simpler example of this sort was given [1898] by René Baire (1874–1932) in a paper that formed part of his doctoral thesis [1899]. Define the sequence of functions f_n on $[0,1]$ as follows: If x is of

15. This follows from the mean value theorem: $f_s(x) = g'(c_s)$ for some c_s between x and $x + h_s$.

the form p/q where p and q are integers with no common divisors and q is less than or equal to n, then $f_n(x) = 0$; otherwise, $f_n(x) = 1$. Then each f_n takes on the value 1 at all but a finite number of points and $|f_n(x)| \leq 1$ for all n and x. The limit function $f(x) = \lim_{n \to \infty} f_n(x)$, however, is totally discontinuous, being equal to 0 if x is rational and to 1 if x is irrational. This function is thus one of Dirichlet's examples of a function that had appeared to him to be unquestionably beyond the purview of the integral concept, and it continued to be regarded in that manner, since it is clearly not Riemann-integrable.

It should be noted in passing that although Baire's work is not directly concerned with integration—the purpose of his example was to show that Dirichlet's function is a "Baire function" of class 2—the profound nature of his results demonstrated what could be accomplished in the theory of functions of a real variable rather than in complex function theory. Undoubtedly Baire's work served to stimulate the interest of mathematicians, including Lebesgue, in the possibilities of research in this field [cf. Lebesgue 1922: 16–18; 1932].

Arzelà had less success with the case in which the f_s are not uniformly bounded. If term-by-term integration is valid, then $G(x) = \lim_{s \to \infty} \int_a^x f_s$ exists and is continuous, since $G(x) = \int_a^x f$. Thus, the existence and continuity of G is a necessary condition for the validity of term-by-term integration. Arzelà thought he could prove the converse of this as well. The reader will observe the close connection between this problem and the question of when a continuous function is an integral (3.3). Indeed, Osgood constructed an example (without a knowledge of Arzelà's paper) of a sequence f_s such that $G(x)$ is one of the continuous monotonic functions with densely distributed intervals of invariability first discovered by Cantor, Scheeffer, and Harnack. In Osgood's example, the set of X-points is uncountable. Osgood was able to prove that when the X-points are countable, Arzelà's theorem is true; Arzelà [1899–1900] then extended this result from continuous f_s to integrable f_s. Although these results are of interest, they are not particularly useful in applications, since the existence of G and its continuity must first be established. By generalizing the condition of uniform boundedness, Lebesgue was to obtain a much more useful result.[16]

16.   We are referring to Lebesgue's well-known "Dominated Convergence Theorem." Both the statement and the proof of this theorem represent simple generalizations of Lebesgue's theorem on uniformly bounded sequences (Section 5.1, Theorem 5.3). The Dominated Convergence Theorem was first published in [Lebesgue 1908: 12].

We have frequently spoken here and in earlier sections of the "problems" and "inadequacies" within Riemann's theory. It should be realized, however, that this characterization of certain developments within that theory is based upon the wisdom of hindsight, upon regarding them from the vantage point of Lebesgue's discoveries. They were not seriously regarded as problems or inadequacies at the time. The nascent theory of functions of a real variable grew out of the development of a more critical attitude, supported by numerous counterexamples, towards the reasoning of earlier mathematicians. Thus, for example, continuous nondifferentiable functions, discontinuous series of continuous functions, and continuous functions that are not piecewise monotonic were discovered. The existence of exceptions came to be accepted and more or less expected. And the examples of nonintegrable derivatives, rectifiable curves for which the classical integral formula is inapplicable, nonintegrable functions that are the limit of integrable functions, Harnack-integrable derivatives for which the Fundamental Theorem II is false, and counterexamples to the classical form of Fubini's Theorem appear to have been received in this frame of mind. The idea, as Schoenflies put it in his report [1900: 111–12], was to proceed, as in human pathology, to discover as many exceptional phenomena as possible in order to determine the laws according to which they could be classified.

But the work of Lebesgue's predecessors should not be discredited. The search for counterexamples to illuminate the logical relationships between various properties certainly plays an important role in modern mathematics. The earlier work, however, also points out the originality and imaginativeness with which Lebesgue viewed the same developments.

5

The Creation of
Modern Integration Theory

5.1   LEBESGUE'S THEORY OF INTEGRATION

Henri Lebesgue (1875–1941) attended the École Normale Supérieure, as had Borel and Baire, and completed his studies there in 1897.[1] The next two years, while he worked at the library of the École, was a productive time for Lebesgue, resulting in the publication of a number of papers. Also during this period, Borel's *Leçons* appeared with its new approach to the measure of sets, and Baire's first papers on discontinuous functions were published. Lebesgue's interest in Baire's work is evident in his second paper [1899a], which is devoted to extending Baire's results to functions of several real variables. In 1899 Lebesgue took a teaching position at the Lycée Central in Nancy, where he remained until 1902. Between June of 1899 and April of 1901, the Academy of Science published a series of five research announcements by Lebesgue in the *Comptes Rendus* [1899b, 1899c, 1900a, 1900b, 1901], which subsequently formed the basis for his doctoral thesis at the Sorbonne; it is in the fifth paper that Lebesgue first announced his generalization of the Riemann integral.

The second paper in this series deals with what he termed the problem of the measure of surfaces; for the class of "surfaces" in the plane bounded by simple, closed curves, the problem was posed in the following manner: "Associate with each surface a number called *area* in such a manner that two equal surfaces have equal areas and that the surface formed by the union of a finite or infinite number of surfaces having portions of common boundary and not overlapping one another has for

1.   For further biographical information on Lebesgue, see Kenneth O. May's biographical sketch and bibliography [Lebesgue 1966: 1–7].

120

its area the sum of the areas of the component surfaces . . ." [1899c: 870]. Apparently, Lebesgue had accepted Borel's view that a measure should be countably additive and had been influenced by Borel and Drach to take the postulational approach. In this connection he was also influenced by Jacques Hadamard, who had proposed the analogous problem of area for polygons, positing, however, only finite additivity [1898: note D]. Hadamard showed that a finitely additive area function defined for polygons must agree with the traditional definition of the area of a polygon.

Lebesgue observed that the interior of a region D bounded by a simple closed curve can be expressed as the sum of a countable number of nonoverlapping polygons, the sum of whose areas is $c_i(D)$. (This is a simple consequence of the definition of $c_i(D)$.) Thus, the postulate of countable additivity leads to defining the area of D as $c_i(D)$. But if D_1 and D_2 have a boundary arc C in common, then

$$c_i(D_1 \cup D_2) = c_i(D_1) + c_i(D_2) + c_e(C),$$

since C becomes part of the interior of $D_1 \cup D_2$. Thus, unless $c_e(C) = 0$ (that is, unless only Jordan-measurable regions are considered) the problem is not solvable: it is impossible to define an additive area function. The postulate of countable additivity does not come into conflict here with Jordan's definition of measurability because Lebesgue restricted himself to unions $D = \bigcup_{n=1}^{\infty} D_n$ such that D belongs to the same class as the D_n. Thus by hypothesis, D is also Jordan-measurable, and therefore $c(D) = \sum_{n=1}^{\infty} c(D_n)$. Lebesgue had not yet found a use for Borel's definition of measure. In the remainder of this paper [1899c], Lebesgue dealt with the analogous problem of area for surfaces in three-dimensional space and then, in a note, gave a definition of the surface integral $\int_S f(x,y)dS$ based on his previous definition of surface area.

It was only after some hesitation that Lebesgue presented the results in the announcements as his doctoral thesis, for he was still unsure of the value of his work, primarily concerned as it was with "pathological" functions [1922: 12–14]. (The mathematicians on the thesis committee were Picard and Goursat.) Certainly he was not encouraged by the fact that the first paper in the series—though it was defended by Picard, who presented all five papers to the Académie—was opposed by Hermite for inclusion in the *Comptes Rendes*. As Lebesgue pointed out,

Hermite's objection came a few years after he had written to Stieltjes "I turn away with fright and horror from this lamentable plague of functions which do not have derivatives" [Hermite and Stieltjes 1905: II, 318].

Lebesgue presented his thesis under the title "Intégrale, longueur, aire," and it was published in the Italian journal *Annali di Matematica* in 1902. The first chapter deals with the measure of sets. Imitating the approach of his earlier note, he posed the problem of defining $m(E)$, a nonnegative measure on bounded sets E, such that:

(1) $m(E) \neq 0$ *for some set E;*

(2) $m(E + a) = m(E)$ *for every real number a ;*[2]

(3)   *If E_n are pairwise disjoint for $n = 1, 2, 3, \cdots$,*   *then*

$$m\left(\bigcup_{n=1}^{\infty} E_n \right) = \sum_{n=1}^{\infty} m(E_n).$$

These properties then imply that $m([0,1]) \neq 0$; if, by convention, the measure of $[0, 1]$ is taken to be 1, it follows that $m\left(\bigcup_{n=1}^{\infty} I_n \right) = \sum_{n=1}^{\infty} L(I_n)$ when the I_n are nonoverlapping intervals. Therefore, if $m(E)$ can be defined, its value will be less than or equal to the numbers

(4) $\sum_{n=1}^{\infty} L(I_n),$

where the I_n denote intervals such that E is contained in their union.[3] The greatest lower bound of the numbers (4) is defined to be the outer measure of E, denoted by $m_e(E)$. Thus, if $m(E)$ is definable, $m(E) \leq m_e(E)$.

Lebesgue's definition of outer measure is a straightforward generalization—guided by Borel's ideas—of the definition of outer content. But Lebesgue's definitions of measurability and inner measure are subtler generalizations of Jordan's ideas. The analogous generalization of inner content is, in fact, no generalization at all; it is equivalent to the definition of inner content. On the other hand, a Jordan-measurable set E has the property that if $E \subset [a,b]$, then by the finite additivity of c_e on measurable sets, $c_e(E) + c_e([a,b] - E) = b - a$. Lebesgue was thus

2. $E + a$ denotes the set of all numbers of the form $x + a$, where x belongs to E.

3. If E is a subset of F, then $m(E) \leq m(F)$; this follows from (3) and the positivity of measure.

probably led to consider the class of bounded sets E such that if $E \subset [a,b]$, then

(5) $$m_e(E) + m_e([a,b] - E) = b - a.$$

If the inner measure of $E \subset [a,b]$ is defined by

$$m_i(E) = b - a - m_e([a,b] - E),$$

then (5) holds if and only if $m_i(E) = m_e(E)$. Bounded sets E such that $m_i(E) = m_e(E)$ were defined by Lebesgue to be measurable.

Lebesgue's definition of measurability can be cast in a slightly different light by emphasizing, as Lebesgue did, its connection with the problem of defining a measure $m(E)$ that satisfies (1)–(3) and also the property $m([0,1]) = 1$. As already noted, if $m(E)$ can be defined, then $m(E) \leq m_e(E)$. Furthermore,

$$m_i(E) = b - a - m_e([a,b] - E) \leq b - a - m([a,b] - E) = m(E),$$

so $m_i(E) \leq m(E) \leq m_e(E)$. For measurable sets at least, there is consequently one and only one solution to the problem of measure, namely $m(E) = m_i(E) = m_e(E)$. Lebesgue's measurable sets therefore represent an extensive class of sets for which the problem of measure, as posed by Lebesgue, has a unique solution.[4]

Next Lebesgue showed that a countable union of measurable sets is also measurable and that (3) holds. Thus the problem of measure is resolved at least for the class of measurable sets. He also derived from (3) the following useful properties:

(6) $$m(E_1 - E_2) = m(E_1) - m(E_2), \quad where \ E_2 \subset E_1;$$

(7) $$m\left(\bigcup_{n=1}^{\infty} E_n \right) = \lim_{n \to \infty} m(E_n), \quad where \ E_n \subset E_{n+1};$$

(8) $$m\left(\bigcap_{n=1}^{\infty} E_n \right) = \lim_{n \to \infty} m(E_n), \quad where \ E_{n+1} \subset E_n.$$

These properties had been noted by Borel (and (7), also by Osgood).
The evident inequality

(9) $$c_i(E) \leq m_i(E) \leq m_e(E) \leq c_e(E)$$

enabled Lebesgue to conclude that all Jordan-measurable sets are Lebesgue-measurable and that the assigned measures are the same. Lebesgue then turned to Borel's definition: Let E denote a Lebesgue-

---

4. Examples of nonmeasurable sets were later constructed by G. Vitali [1905b] and E. B. Van Vleck [1908], but Lebesgue [1907c] had his reservations about such constructions.

measurable set and ϵ_n any sequence of numbers which decreases monotonically to 0. By definition, E can be enclosed in a countable number
of nonoverlapping intervals whose union A_i has measure $m(E) + \epsilon_i$.

Then $E_1 = \bigcap\limits_{i=1}^{\infty} A_i$ is Borel-measurable and contains E, and $E_1 - E$ has

zero measure. Thus, $E_1 - E$ can be enclosed in a countable union B_i of
intervals such that $m(B_i) = \epsilon_i$ and B_i is contained in A_i. The set

$e = \bigcap\limits_{i=1}^{\infty} B_i$ is consequently Borel-measurable and its measure is 0. In sum-

mary, Lebesgue had proved the following theorem (with $E_2 = E_1 - e$):

> THEOREM 5.1.   *If E is measurable, then there exist Borel mea-
> surable sets E_1 and E_2 such that $E_1 \supset E \supset E_2$ and $m(E_2) = m(E) = m(E_1)$.*

This result shows, as Lebesgue pointed out, that Lebesgue-measurable
sets are precisely those sets assigned a definite measure by the "calculus
of inequalities" that Borel had introduced for non-Borel sets. Lebesgue
was to make effective use of this relationship.

Lebesgue's theory of measure, although it was developed with a
much greater degree of clarity and generality than Borel's, represents a
natural extension of Borel's ideas. The credit for applying this theory
of measure to integration theory, however, goes to Lebesgue. The application is made in the second chapter of his thesis, which begins with a
discussion of the relationship between the content of sets and the integral (along the lines indicated in the Introduction):

$$(10) \qquad \overline{\int_a^b} f = c_e(E^+) - c_i(E^-) \quad \text{and} \quad \underline{\int_a^b} f = c_i(E^+) - c_e(E^-),$$

where E^+ and E^- denote the regions bounded by the graph of f and lying
above and below the x-axis, respectively. Although previously these
relationships had not been stated with this degree of precision and
generality, they were recognized by Harnack, Peano, and undoubtedly
others as well. When regarded by Lebesgue these relationships "immediately suggest the following generalization . . ." [1902a: 250]: Let f
be defined and bounded on $[a,b]$. If the region $E = E^+ \cup E^-$ bounded by
the graph of f is measurable (so E^+ and E^- are also measurable), an
integral can be defined by

$$(11) \qquad \int_a^b f = m(E^+) - m(E^-).$$

Similarly, upper and lower integrals can be defined as in (10). The relationship given by (11) represents the geometrical version of Lebesgue's definition of the integral. Because of the inequalities in (9) it is at once clear that Lebesgue's definition includes Riemann's definition as a special case. But this fairly straightforward generalization marked only the beginning of Lebesgue's accomplishment, for in his thesis and subsequent works he brilliantly demonstrated the superiority of his definition over those of his predecessors.

The first step in this direction was to obtain a more workable formulation of the generalized integral, that is, to define the integral analytically, similar to the manner in which the Riemann integral is defined as the limit of Cauchy sums. Suppose for simplicity that f is a nonnegative function defined on $[a,b]$. The analytical form of Riemann's definition is obtained by summing the areas of the inscribed and of the circumscribed rectangles that approximate the area of $E = E^+$. Lebesgue's idea was to partition the range of f instead of the domain, $[a,b]$. In other words, if m and M denote the greatest lower bound and the least upper bound of f on $[a,b]$, let P denote the partition

$$m = a_0 < a_1 < a_2 < \cdots < a_n = M$$

of the interval $[m,M]$, and consider the set e_i of all x such that

(12) $$a_i \leq f(x) < a_{i+1}.$$

Then $e_0, e_1, e_2, \cdots, e_{n-1}$ partition $[a,b]$, and the planar set E lies between the generalized rectangles with base e_i and heights a_i and a_{i+1}, respectively (see Figure 6). The measures of these two sets are $\sum_{i=0}^{n-1} a_i m(e_i)$ and $\sum_{i=0}^{n-1} a_{i+1} m(e_i)$. In other words, $\int_a^b f = m(E)$ lies between these two measures. But their difference is $\sum_{i=0}^{n-1} (a_{i+1} - a_i) m(e_i) \leq \|P\|(b-a)$, where $\|P\|$ denotes as before the maximum of the differences $a_{i+1} - a_i$. Consequently, $\int_a^b f = \lim_{\|P\| \to 0} \sum_{i=0}^{n-1} a_i m(e_i) = \lim_{\|P\| \to 0} \sum_{i=0}^{n-1} a_{i+1} m(e_i)$.

For this reasoning to be complete, the sets e_i must be measurable. The measurability of the e_i follows from the proposition that if f has an integral by (11), then, for any real number a, the set of points x such that $f(x) > a$ is measurable. Lebesgue established this proposition by using Theorem 5.1 and the fact that the intersection of a planar Borel-measurable set with a line parallel to one of the coordinate axes is a linear Borel-measurable set. If, for a function f (bounded or not), the set of x such that $c < f(x) < d$ is measurable for all values of c and d, then f was termed a "summable function." We shall follow Lebesgue's later usage and refer to such functions as "measurable functions." Lebesgue

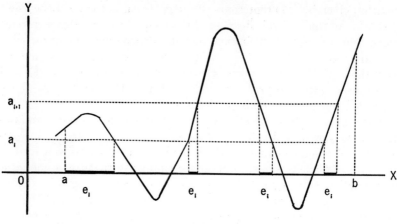

Fig. 6

thus showed that his generalization of the integral can be expressed in the following manner: If f is defined, bounded, and measurable on $[a,b]$, then f is said to be summable and the integral of f is

$$(13) \qquad \int_a^b f = \lim_{\|P\|\to 0} \sum_{i=0}^{n-1} a_i m(e_i) = \lim_{\|P\|\to 0} \sum_{i=0}^{n-1} a_{i+1} m(e_i),$$

where e_i consists of those x satisfying (12).

Lebesgue's generalization of the integral should be compared with the "natural" generalization of Jordan's definition of the Riemann integral (as discussed in the Introduction). From a purely logical standpoint, the two generalizations are equivalent,[5] and undoubtedly Jordan's reformulation of Riemann's definition motivated Lebesgue to consider the possibility of an analytical definition based upon partitioning $[a,b]$ into (Lebesgue) measurable sets instead of into just intervals. From a historical point of view, however, it seems extremely important that Lebesgue used a partition of $[m,M]$ to determine the nature of the sets e_i into which $[a,b]$ is partitioned. Such an approach necessitated the introduction of the notion of a measurable function; and it was the properties of measurable functions and the structure of the sets e_i that guided Lebesgue's reasoning and led to his major results. (The importance of these notions is especially apparent in the proofs of Theorems 5.2, 5.3, 5.8, 5.10, 5.11, and 5.12.) That the form of Lebesgue's definition was significant is perhaps confirmed by the fact that W. H.

5. See, for example, J. H. Williamson, *Lebesgue Integration*, New York, 1962, pp. 39ff, especially Theorem 3.3e and Corollary 1 on pp. 50–51.

Young, in generalizing Jordan's definition by the "natural" generalization, did not discover any of the beautiful theorems that mark Lebesgue's achievement (see 5.2).

By means of the analytical definition, Lebesgue was able to establish the usual properties of the integral. The distinction between the old and new definitions, however, began to be revealed with the following proposition:

> THEOREM 5.2.   *If f_n is a sequence of measurable functions defined on $[a,b]$ and $f(x) = \lim_{n \to \infty} f_n(x)$ exists, then f is also measurable.*

The proof is a simple consequence of the properties of measurable sets.

Baire's work had called attention to the possibility of classes of functions with the property that the "limit functions," $f(x) = \lim_{n \to \infty} f_n(x)$, of functions f_n in the class likewise belong to the class (see 4.4). Baire's research led him to the idea of classifying functions in the following manner (all functions discussed are assumed to be defined on a common set of points): A function f is defined to be of class 0 if f is continuous; of class 1 if f is not continuous but $f(x) = \lim_{n \to \infty} f_n(x)$, where the f_n are continuous; of class 2 if f is not of class 0 or 1 but $f(x) = \lim_{n \to \infty} f_n(x)$, where the f_n are of class 0 or 1. By means of transfinite induction, it is possible to define functions of class γ for each ordinal number less than Ω. Baire proved that E, the class consisting of all functions of class γ for some γ less than Ω, contains all its limit functions.

Theorem 5.3 may have been suggested to Lebesgue by Baire's result about the class E. The theorem shows that the "Baire functions"—that is, the members of E—are always measurable and thereby reveals, as Lebesgue noted, the broader scope of his theory of integration. In particular, Dirichlet's celebrated example of a function that appeared intractable in terms of the integral concept—the function that takes the value 1 when x is rational and the value 0 when x is irrational—possesses a Lebesgue integral since, as Baire had observed, the function is of class 2. In fact at this time Lebesgue was unable to construct an example of a bounded, nonsummable function because he could not construct a nonmeasurable set.

Another method of constructing summable functions that are not Riemann-integrable was discovered by Lebesgue. Lebesgue used a proposition that had evaded his predecessors: A bounded function is integrable if and only if its points of discontinuity form a set of measure zero. If f and g are continuous and $g(x) \neq 0$, let the function F be defined by $F(x) = f(x)$ for x in S and $F(x) = f(x) + g(x)$ for x in $[a,b] - S$, where S is such that $[a,b] - S$ is dense in $[a,b]$ but has zero measure. Then F cannot be integrable, since it is everywhere discontinuous.

The reason why F is summable illustrates an important characteristic of Lebesgue's analytical definition (13): it yields a definition of $\int_E f$ for any bounded measurable set E; all that is required is to replace $[a,b]$ by E. Furthermore, if f is summable with respect to two disjoint measurable sets E_1 and E_2, f is clearly summable on their union. F is summable on S (F is identical with f on S) and on the complement of S (since the complement of S has measure zero); hence, F is summable on $[a,b]$, the union of S and its complement. Also implied by this reasoning is the fact that the summability of a function and the value of its integral are not affected by altering or ignoring the values of the function on a set of zero measure, a fact that was later exploited by Fubini (5.3).

One of Lebesgue's most useful results is the following extension of the theorems of Arzelà and Osgood (4.4):

THEOREM 5.3. *If f_n is a sequence of measurable functions defined on a measurable set E such that $\left| f_n(x) \right| \leq B$ for all x in E and all n, and if $f(x) = \lim_{n \to \infty} f_n(x)$ exists, then $\int_E f = \lim_{n \to \infty} \int_E f_n$.*

The proof is elegantly simple and must be compared with the proofs by Arzelà and Osgood. Let ϵ denote a fixed positive number and let e_n denote the set of all x in E for which $\left| f_{n+p}(x) - f(x) \right| \geq \epsilon$ for at least one value of p. Then e_n is measurable, e_{n+1} is contained in e_n, and $\bigcap_{n=1}^{\infty} e_n = \varnothing$. Thus by (8), $\lim_{n \to \infty} m(e_n) = 0$. The theorem then follows from the inequalities

$$\left| \int_E (f - f_n) \right| \leq \int_{e_n} \left| f - f_n \right| + \int_{E - e_n} \left| f - f_n \right| \leq 2Bm(e_n) + \epsilon m(E).$$

When he composed his thesis, Lebesgue was apparently not familiar with Arzelà's work, but Osgood's results were known to him and clearly motivated Theorem 5.3. The idea behind the proof of Theorem 5.3—to consider the sets e_n—may also have been suggested to Lebesgue by Osgood's Lemma and the role it plays in Osgood's proof of his analogous theorem.

The discoveries of Dini, Volterra, and others had indicated that the

Fundamental Theorem II, viz. $\int_a^b f' = f(b) - f(a)$, loses its general validity in Riemann's theory because of the existence of functions with bounded, nonintegrable derivatives. Lebesgue was familiar with the literature discussing the existence of such functions, and he saw that Theorem 5.3 restores the Fundamental Theorem II, at least for bounded derivatives. Specifically, he proved:

THEOREM 5.4. *If f' exists and is bounded on $[a,b]$, then f' is summable and $\int_a^b f' = f(b) - f(a)$.*

The proof is based on the fact that the functions $g_h(x) = \dfrac{f(x+h) - f(x)}{h}$ satisfy the conditions of Theorem 5.3. Thus, since the continuity of f ensures that $\left(\int_a^x f \right)' = f(x)$,

$$\int_a^b f' = \lim_{h \to 0} \int_a^b g_h = \left[\lim_{h \to 0} \frac{1}{h} \int_x^{x+h} f \right]_a^b = f(b) - f(a).$$

Prior to Lebesgue, several attempts had been made to extend the integral concept to unbounded functions, some of which were discussed in 3.3. The analytical form (13) of Lebesgue's definition afforded him the means of obtaining a natural and much more useful extension of his theory to unbounded functions; for if f is defined and measurable on $[a,b]$, one can consider a partition P of $(-\infty, +\infty)$, viz.

$$\cdots < a_{-2} < a_{-1} < a_0 < a_1 < a_2 < \cdots,$$

where $\|P\|$ is finite. As before, the set e_i is defined to consist of those x such that $a_i \leq f(x) < a_{i+1}$. If the series $\sum_{i=-\infty}^{\infty} a_i m(e_i)$ converges absolutely, so does $\sum_{i=-\infty}^{\infty} a_{i+1} m(e_i)$, and conversely. When the series so converge for some partition P, the function f is called "summable," because as $\|P\| \to 0$ both sums approach a common limit, which is defined to be $\int_a^b f$. Clearly f is summable if and only if $|f|$ is summable. (It should be noted that Lebesgue's definition completely disregards values of x such that $f(x)$ is infinite, and he assumed that $f(x)$ will be infinite at no more than a finite number of points.) It follows directly from this definition that:

THEOREM 5.5. *If f is summable on $[a,b]$ and $\epsilon > 0$ is given, then there exists a measurable subset E of $[a,b]$ such that f is bounded on E and* $\displaystyle\int_{E^c} |f| < \epsilon$, *where* $E^c = [a,b] - E$.

The proposition enabled Lebesgue to extend the basic properties of the integral of bounded functions to properties of unbounded functions. In particular, the continuity of $\displaystyle\int_a^x f$ is established in this manner.

Next Lebesgue considered the Fundamental Theorem II when f' is unbounded. Harnack had encountered difficulties in this connection because of the existence of monotonic continuous functions with densely distributed intervals of invariability (3.3). Lebesgue was able to obtain results of a more positive nature with his definition. When a function f possesses a finite-valued derivative f', then, by Theorem 5.2, f' is automatically measurable. But f' need not be summable, a fact that is illustrated by the function f such that $f(0) = 0$ and $f(x) = x^2 \sin(1/x^2)$ for $x \neq 0$. On the other hand, Lebesgue was able to establish the following important result:

THEOREM 5.6. *If f' is finite-valued, then f' is summable if and only if f is of bounded variation; and when f' is summable,*
$$\int_a^b f' = f(b) - f(a).$$

The idea behind the proof is to show that the total variation of f on $[a,b]$, be it finite or infinite, is precisely $\displaystyle\int_a^b |f'|$ where this is taken to be infinite when f' is not summable.[6] A slight modification in the reasoning yields:

THEOREM 5.7. *If C is a continuous curve defined by $x = x(t)$, $y = y(t)$, $z = z(t)$ with the property that x', y', and z' exist and are finite-valued, then $\displaystyle\int_a^b [(x')^2 + (y')^2 + (z')^2]^{1/2}$ exists if and only if C is rectifiable; and when C is rectifiable, its length is given by this integral.*

This result should be considered in the light of the developments regarding the relationship of the concepts of integral and curve length

6. Jordan had proved in his *Cours* [1882–87] that $G(x) = \int_a^x g$ is of bounded variation when g is Riemann-integrable and that the total variation of G on $[a,b]$ is at most $\int_a^b |g|$.

(see 3.4). Lebesgue was familiar with du Bois-Reymond's papers on this subject as well as the later literature and considered Theorem 5.7 to be in accord with duBois-Reymond's approach to the concept of curve length [Lebesgue 1902a: 234]. Thus, the developments discussed in 3.4 motivated Theorem 5.7, which in turn may have motivated the purely analytical Theorem 5.6. Lebesgue's interest in the theory of curve rectification—in particular, in the question of the validity of the classical integral formula for curve length—was closely connected with his discovery that any continuous function of bounded variation possesses a finite derivative except on a set of measure zero.

Lebesgue's chapter on integration concludes with some reflections: In order that a generalization be useful it must continue to satisfy the basic conditions characteristic of the integral, including the condition that "integration permit the resolution of the fundamental problem of the integral calculus: to find a function knowing its derivative . . ." [1902a: 281]. At least for bounded functions, Lebesgue noted, his definition, unlike Riemann's, resolves this problem. "Not being able to demonstrate that the definition proposed was the only one fulfilling the conditions, I have attempted to show that it was natural and that from a geometrical point of view it seemed almost necessary" [1902a: 282].

There is no doubt, as J. C. Burkill remarked [1942–44: 484], that Lebesgue's thesis represents "one of the finest which any mathematician has ever written." Nevertheless, there were certain properties of the Riemann integral that Lebesgue was unable to establish for his integral. The most notable of these is the property expressed by the Fundamental Theorem I, namely, that $\left(\int_a^x f \right)' = f(x)$ "in general" in Harnack's terminology (see 3.1). That property of the Riemann integral was derived from the fact that an integrable integrand f must be continuous in general; one would not expect a similar property to hold for a definition of the integral that admits totally discontinuous functions as integrands. It is consequently not surprising that we find no discussion of the Fundamental Theorem I in Lebesgue's thesis. Lebesgue also ran into difficulties trying to imitate Dini's results on the Fundamental Theorem II for the four Dini derivatives (2.3). In the first place, it is not immediately clear that these derivatives are measurable, although Lebesgue was able to establish this much. But he was only able to derive the inequality

$$\int_a^b D_+ f \leq f(b) - f(a) \leq \int_a^b D^+ f$$

when the Dini derivatives are bounded and, hence, summable.

Lebesgue was chosen to give the Cours Peccot at the Collège de France during the academic year 1902–03—a privilege that had belonged to Borel from 1899 to 1902. It turned out to be a fruitful period for Lebesgue, and he was able to resolve many of the questions left open in his thesis and to make further discoveries [1903b, 1903c]. His lectures, which contained these newly acquired results, were published in 1904 under the title *Leçons sur l'intégration et la recherche des fonctions primitives* as the sixth volume in a series of monographs on the theory of functions edited by Borel. (It should be mentioned that, according to Lebesgue [1922: 14], Borel was the first mathematician—including Lebesgue himself—to be convinced of the value of his work on the theory of integration.) The *Leçons* [1904] fully documents Lebesgue's familiarity with the literature discussed in the previous chapters and will serve as the basis for our discussion of his new results.

The last chapter of the *Leçons* is devoted to Lebesgue's theory of integration. Here he began, again imitating Drach and Borel, by presenting the "problem of integration" as the problem of defining a real number $\int_a^b f$ corresponding to f and the interval $[a,b]$ such that the following six properties hold:[7]

$$(14) \qquad \int_a^b f(x)dx = \int_{a+h}^{b+h} f(x-h)dx;$$

$$(15) \qquad \int_a^b f + \int_b^c f + \int_c^a f = 0;$$

$$(16) \qquad \int_a^b (f+g) = \int_a^b f + \int_a^b g;$$

$$(17) \qquad \textit{If } f(x) \geq 0 \textit{ and } b > a, \textit{ then } \int_a^b f \geq 0;$$

$$(18) \qquad \int_0^1 1 = 1;$$

$$(19) \qquad \textit{If } f_n(x) \nearrow f(x), \textit{ then } \int_a^b f_n \nearrow \int_a^b f.$$

Lebesgue was then able to clarify the question of the necessity of his

---

7. This type of characterization of the integral played an important role in the definition of the integral formulated by P. J. Daniell [1917–18].

definition that he had raised in his thesis. In the first place, (16) and (17) imply that, for any real number k,

$$(20) \qquad \int_a^b kf = k \int_a^b f,$$

while (14), (15), and (16) show that

$$(21) \qquad \int_a^b 1 = b - a.$$

Because of (21), it is not difficult to show that if $f = \chi_E$ where E is a measurable subset of $[a,b]$, then the only possible definition of $\int_a^b f$ satisfying (14)–(19) is

$$(22) \qquad \int_a^b \chi_E = m(E).$$

(For example, using (16), (17), (19), and (21) one can conclude that $m_i(E) \le \int_a^b \chi_E \le m_e(E)$.) Once (22) has been established, it follows easily that when f is bounded and measurable, the only possibility for $\int_a^b f$ satisfying (14)–(17) is the Lebesgue integral: for if

$$(23) \qquad \varphi = \sum_{i=0}^{n-1} a_i \chi_{e_i} \quad \text{and} \quad \Phi = \sum_{i=0}^{n-1} a_{i+1} \chi_{e_i},$$

where the sets e_i and the partition $\{a_i\}$ are defined as in Lebesgue's definition of the integral (equations (12) and (13)), then (14)–(22) necessitate that

$$\int_a^b \varphi = \sum_{i=0}^{n-1} a_i m(e_i) \quad \text{and} \quad \int_a^b \Phi = \sum_{i=0}^{n-1} a_{i+1} m(e_i).$$

Both of these sums, of course, converge to the Lebesgue integral of f as $\|P\| \to 0$, and so (17) applied to the inequalities

$$(24) \qquad \varphi(x) \le f(x) \le \Phi(x), \qquad 0 \le \Phi(x) - \varphi(x) \le \|P\|$$

leaves no doubt as to the value of $\int_a^b f$. Similar considerations show that, more generally, if f is any summable function, then the only definition of $\int_a^b f$ satisfying (14)–(19) is the Lebesgue integral. Thus "the problem of integration is possible and in a single way if it is posed for summable functions" [1904: 115].

In the *Leçons* Lebesgue was able to present further results on the Fundamental Theorem. Because his proofs were highly condensed and partially inaccurate, some of the results were questioned by the Italian mathematician Beppo Levi [Levi 1906a, 1906b, 1906c]. In answer to Levi, Lebesgue amplified and corrected his original arguments—and at the same time made his ideas better known to Italian mathematicians. We shall refer to these emendations when we sketch the basically sound proofs presented in the *Leçons*.

In his thesis, Lebesgue had been unable to extend the Fundamental Theorem II (Lebesgue's Theorem 5.6) to Dini derivatives. By introducing what he called a chain of intervals [1904: 61–63], he was able both to simplify his original proof of Theorem 5.6 and to extend it to Dini derivatives; the theorem becomes:

> THEOREM 5.8.   *Let f be continuous on [a,b] and let Df denote one of the four Dini derivatives of f. Then if Df is finite-valued, Df is summable if and only if f is of bounded variation. When Df is summable,* $\int_a^b Df = f(b) - f(a)$.

Basically, the method of proof is the same as for Theorem 5.6: to determine a sum that simultaneously approximates not only the total variation of f on $[a,b]$ but also $\int_a^b |Df|$. The proof of Theorem 5.8, however, uses chains of intervals. The following proposition defines and establishes the existence of such a chain:

> THEOREM 5.9.   *Suppose that to every x in [a,b) there corresponds an interval [x,x+h], h>0, contained in [a,b]. Then there exist a countable number of nonoverlapping intervals $[x_i,x_i+h_i]$ whose union is [a,b]. These intervals are said to form a chain of intervals from a to b.*

The chain is defined inductively: Let $[a,x_1]$ be the interval corresponding by hypothesis to $x=a$; $[x_1,x_2]$, the interval corresponding to $x=x_1$; and continue in this manner. If, for every integer n, x_n is less than b, then x_n is an increasing sequence that therefore converges to a limit, $x_\omega \leq b$. If $x_\omega < b$, let $[x_\omega,x_{\omega+1}]$ denote the interval corresponding to $x=x_\omega$, and so on by transfinite induction. After a countable number of steps, there must be some x_α such that $x_\alpha = b$, since for any integer n there can be only a finite number of intervals of length greater than $1/n$.

The proof of Theorem 5.8 for D^+f is then carried out in the following manner [1904: 121–123]: Let a positive number ϵ be fixed and let P denote any partition $\cdots < L_{-1} < L_0 < L_1 < \cdots$ of $(-\infty, \infty)$ with $\|P\| < \epsilon$. The set e_n is defined to consist of all x in $[a,b]$ such that

$$L_n < D^+f(x) \le L_{n+1},$$

and the e_n are arranged in a simply ordered sequence $e_{n_1}, e_{n_2}, e_{n_3}, \cdots$. A positive sequence a_{n_i} is chosen so that

(25)
$$\sum_{i=1}^{\infty} |L_{n_i}| \, a_{n_i} < \epsilon.$$

Since e_{n_1} is measurable, it can be enclosed in a union of nonoverlapping intervals A_{n_1}; similarly, $[a,b] - e_{n_1}$ can be enclosed in a union I_{n_1} such that $m(A_{n_1} \cap I_{n_1}) < a_{n_1}$. In general, e_{n_k} and $[a,b] - (e_{n_1} \cup e_{n_2} \cup \cdots \cup e_{n_k})$ can be enclosed in unions of nonoverlapping intervals A_{n_k} and I_{n_k}, respectively, such that $m(A_{n_k} \cap I_{n_k}) < a_{n_k}$ and A_{n_k} and I_{n_k} are both contained in $I_{n_{k-1}}$. It therefore follows that

(26)
$$m(A_{n_k}) - m(e_{n_k}) < a_{n_k}.$$

A chain of intervals is now determined by associating with each x in $[a,b)$ the largest positive h such that $h \le \epsilon$ and such that:

(27) $(x, x+h)$ *is contained in* A_p, *where* x *is in* e_p;

(28)
$$L_p \le \frac{f(x+h) - f(x)}{h} \le L_{p+1} + \epsilon.\text{[8]}$$

Let $[x_i, x_i + h_i]$ denote a chain from a to b in accordance with Theorem 5.9, and let B_p denote the union of all those intervals $[x_i, x_i + h_i]$ for which x_i belongs to e_p. Lebesgue then introduced the expressions

$$v_1 = \sum_p |L_p| \, m(B_p) \quad \text{and} \quad v = \sum |f(x_i + h_i) - f(x_i)|,$$

where the summation for v is over the entire chain. Lebesgue had proved earlier that v can be made arbitrarily close to the total variation of f on $[a,b]$ by taking ϵ small enough. Similarly, the choice of the sets A_{n_i} and I_{n_i} satisfying (25) and (26) was intended to guarantee that v_1 can be made arbitrarily close to $\sum |L_p| \, m(e_p)$—and hence to $\int_a^b |D^+f|$—for

8. Lebesgue's notation is perhaps somewhat confusing here. A_p denotes that A_{n_i}, where $p \leftrightarrow n_i$ in the reordering of the sets e_p into a simple sequence.

ϵ sufficiently small and that $\sum\limits_{p=-\infty}^{\infty} L_p m(B_p)$ is likewise close to $\int_a^b D^+f$ when D^+f is summable. This will be the case if the sequence a_{n_i} is chosen a bit more carefully, as Lebesgue later showed [1906b] after Beppo Levi's criticism.

The proof of the theorem now follows by the observation that (28) implies the inequalities

$$
\begin{align}
(29) \quad & L_p h_i \leq f(x_i + h_i) - f(x_i) \leq (L_p + 2\epsilon)h_i; \\
& (\,|\,L_p\,| - 2\epsilon)h_i \leq |\,f(x_i + h_i) - f(x_i)\,| \leq (\,|\,L_p\,| + 2\epsilon)h_i.
\end{align}
$$

The summation of these inequalities over all i such that x_i is in e_p, for p fixed, and then over all p, yields:

$$
\begin{align}
(30) \quad & \sum L_p m(B_p) \leq f(b) - f(a) \leq \sum L_p m(B_p) + 2\epsilon(b - a); \\
& v_1 - 2\epsilon(b - a) \leq v \leq v_1 + 2\epsilon(b - a).
\end{align}
$$

From (29) and (30) Lebesgue therefore concluded that

$$
\int_a^b D^+f = f(b) - f(a)
$$

when D^+f is summable and that this is the case if and only if f is of bounded variation.

With this result, Lebesgue matched what Dini had accomplished within the context of Riemann's theory (2.3) and obtained a necessary and sufficient condition for the Fundamental Theorem II in the case of unbounded derivatives, whereas Harnack had only been able to recognize the difficulties that arise (3.3). An immediate consequence of Lebesgue's result is: If f is a continuous function of bounded variation such that all four of its Dini derivatives are finite-valued, then $f'(x)$ exists as a finite number except on a set of measure zero. This consequence comes from the fact that all four Dini derivatives have the same integral over any interval. In fact,

$$
f(b) - f(a) = {}^* \!\int_a^b f',
$$

where * signifies integration over all x in $[a,b]$ for which $f'(x)$ exists. Lebesgue's treatment of curve length was to enable him to establish the more general result that any continuous function of bounded variation has a finite derivative except on a set of zero measure.

For continuous functions f, there are two cases in which Theorem 5.8 is not applicable: (1) Df is finite-valued but not summable, and (2) Df is not finite-valued. In his thesis [1904: 269ff], Lebesgue briefly considered the problem of extending the Fundamental Theorem II (Theorem 5.6) to nonsummable derivatives and therefore proposed a generalization of his integral analogous to Hölder's (3.3). He showed by example, however, that this mode of definition does not cover all finite-valued derivatives. The year after the publication of *Leçons sur l'intégration*, Hans Hahn [1905] constructed two continuous functions, f_1 and f_2, such that $f_1'(x) = f_2'(x)$ for all x, but f_1 and f_2 do not differ by a constant. Hahn's functions—constructed as one might expect by using a perfect nowhere dense set—have derivatives that are not everywhere finite. Hence, at least for functions f of type (2), the Fundamental Theorem II does not apply, since f is not determined up to a constant by f' and therefore could not be obtained from f' by an integration process.

Hahn's paper prompted Lebesgue [1907b] to reconsider whether the Fundamental Theorem II could be extended to functions of type (1). Although Lebesgue arrived at no positive conclusions, his observations on how to approach the construction of f from f' were later successfully developed by Arnaud Denjoy. While still a student at the École Normale, Denjoy attended Baire's lectures on discontinuous functions, which Baire presented as the Cours Peccot at the Collège de France the year after Lebesgue had presented his lectures on the integral. Also, Denjoy was responsible for preparing Baire's lectures [Baire 1905] for publication in Borel's series of monographs, and thus he became quite familiar with Baire's techniques. By combining Lebesgue's observations with the transfinite methods developed by Baire,[9] Denjoy extended the integral concept to certain nonsummable finite-valued functions, including all finite-valued derivatives Df, and established the Fundamental Theorem II for any function f of type (1).[10]

A more surprising advance than Theorem 5.8 was Lebesgue's discovery that continuity is not essential for the validity of the Fundamental Theorem I:

THEOREM 5.10. *If f is summable on $[a,b]$, then* $\left(\int_a^x f \right)' = f(x)$

almost everywhere.[11]

9. Denjoy's method is completely analogous to that employed by Baire [1899: 33ff] in his thesis. The method used in Baire's *Leçons* [1905] to prove the same result is somewhat different.

10. Denjoy [1912a; 1912b] proved the Fundamental Theorem II for $Df = f'$; later [1916] he extended it to Dini derivatives.

11. In the course of establishing this and related results, Lebesgue [1903c: 1228] discovered that continuity and measurability are actually related in the following sense: If f is measurable, then for almost all values

The phrase "almost everywhere"—analogous to Harnack's "in general"—was later introduced by Lebesgue to signify that a condition holds for all points except those forming a set of measure zero.

The proof of Theorem 5.10 employs a useful technique to which Lebesgue frequently resorted: a result is established first for characteristic functions χ_E, then for linear combinations of such functions, and finally for summable functions in general. The motivation for this approach is, of course, that a linear combination of characteristic functions can be determined whose integral is arbitrarily close in value to that of a given summable function. This technique is similar to du Bois-Reymond's technique of extending the formula of integration by parts to Riemann integrals (2.2). It should be noted that Lebesgue esteemed du Bois-Reymond's work.

Thus Lebesgue began by considering a function of the form $f = \chi_E$: Let ϵ_p denote a sequence of positive numbers decreasing to 0. Then there exists a union of nonoverlapping intervals A_p such that $m(A_p) - m(E) < \epsilon_p$ and $m(E_1) = m(E)$, where E_1 denotes the intersection of the A_p. (See the proof of Theorem 5.1.) Hence, $\int_a^x \chi_E = \int_a^x \chi_{E_1}$, and it suffices to consider the differentiability properties of the latter. If

$$F_p(x) = \int_a^x \chi_{A_p} \quad \text{and} \quad F(x) = \int_a^x \chi_{E_1}$$

then

$$1 \geq \frac{F_p(x + h) - F_p(x)}{h} \geq \frac{F(x + h) - F(x)}{h} \geq 0;$$

therefore,

$$1 \geq D^+F_p(x) \geq D^+F(x) \geq 0.$$

At this point, although he offered no proof, Lebesgue made use of the fact that $D^+F_p(x) = 0$ except for x in A_p (where, clearly, $D^+F_p(x) = 1$) and a set of measure zero. In geometrical terms—and it must be remembered that Lebesgue had a penchant for geometry and that a geometrically oriented point of view pervades his work—Lebesgue assumed that the metric density of A_p is equal to $\chi_{A_p}(x)$ at almost

---

of x an interval (α, β) containing x can be determined corresponding to any pair of positive numbers ϵ_1 and ϵ_2 such that $|f(x') - f(x)| < \epsilon_1$ except for values of x_1' belonging to a set of measure $\epsilon_2 (\beta - \alpha)$. The relationship between measurability and continuity has since been expressed in a number of forms. See, for example, I. P. Natanson, *Theory of Functions of a Real Variable*, Vol. 1, New York, 1955, pp. 101ff.

every point x.[12] It may have been his realization of this geometrically plausible property of a sum of nonoverlapping intervals that led Lebesgue to see the implications for the differentiability of $\int_a^x f$.

The fact that $D^+F(x) = 0$ except for x in A_p and a set of measure zero, together with (30), demonstrates that $D^+F(x) = 0$ except for x in E_1 and a set of measure zero. Moreover, since F and D^+F satisfy the conditions of Theorem 5.8,

$$\int_a^b D^+F = F(b) - F(a) = m(E_1).$$

Therefore, D^+F cannot be strictly less than 1 on a subset of E_1 of positive measure. In other words, $D^+F(x) = \chi_{E_1}(x)$ almost everywhere.

To extend Theorem 5.10 to summable functions, Lebesgue introduced the functions φ and Φ defined by (23). The inequalities in (24) imply that

$$(31) \qquad \frac{1}{h}\int_x^{x+h} \varphi \leq \frac{1}{h}\int_x^{x+h} f \leq \frac{1}{h}\int_x^{x+h} \Phi.$$

Suppose first that f is bounded. Then φ and Φ are finite linear combinations of characteristic functions, so $\left(\int_a^x \varphi\right)' = \varphi(x)$ and $\left(\int_a^x \Phi\right)' = \Phi(x)$ both hold almost everywhere. This result together with (31) immediately shows that $\varphi(x) \leq D^+F(x) \leq \Phi(x)$ almost everywhere, where $F(x) = \int_a^x f$; since $\varphi(x)$ and $\Phi(x)$ converge to $f(x)$ as $\|P\| \to 0$, then $f(x) \leq D^+F(x) \leq f(x)$ almost everywhere, or $D^+F(x) = f(x)$ almost everywhere. The same argument applies to the other Dini derivatives, so Theorem 5.10 is proved when f is bounded.

Lebesgue later considered the case for Theorem 5.10 in which f is not bounded (after Beppo Levi questioned the validity of the theorem in this case), but he assumed that the functions $\int_a^x \varphi$ and $\int_a^x \Phi$ have finite-valued Dini derivatives. The proof for the remaining case, which he did not supply until still later [1907a], is already contained in its

---

12. The metric density of a measurable set E at a point x is $\lim_{\delta \to 0} \dfrac{m(E \cap (x-\delta, x+\delta))}{2\delta}$ when this limit exists. A proof of the above assertion concerning A_p was first published by Lebesgue [1910b: 405] in 1910. The notion of density, however, was used by him as early as 1903 [Lebesgue 1903c: 1229n1].

essentials elsewhere in the *Leçons* [1904: 128]. The proof is based upon the following important result:

THEOREM 5.11.   *If f is continuous and of bounded variation on [a,b], then its Dini derivatives are finite almost everywhere.*

Lebesgue proved Theorem 5.11 by making some adjustments in his proof of Theorem 5.8. These adjustments are worth considering in some detail, because Theorem 5.11 forms a crucial link in the chain of reasoning leading to Lebesgue's discovery that continuous functions of bounded variation are differentiable almost everywhere.

In the proof of Theorem 5.8, the sets e_p had contained all the points of $[a,b]$. For Theorem 5.11, the sets $e_{-\infty}$ and $e_{+\infty}$ must be added, where these are defined respectively as the sets of points where $D^+f(x) = -\infty$ and $D^+f(x) = +\infty$. Associated with these sets are the sets $A_{-\infty}$ and $A_{+\infty}$, analogous to the sets A_p. A chain of intervals from a to b, as in the proof of Theorem 5.8, can still be obtained by associating with each x in $e_{-\infty}$ or $e_{+\infty}$ an interval $[x,x+h]$ defined in the following manner: Let $M = M(\epsilon) \nearrow \infty$ as $\epsilon \searrow 0$, where ϵ denotes the ϵ of Theorem 5.8, and let h denote the maximum positive number not exceeding ϵ such that:

(27') $(x, x + h)$ *is contained in* $A_{-\infty}$ *(or* $A_{+\infty}$*)*;

(28') $\dfrac{f(x + h) - f(x)}{h} < -M \text{ (or } > M).$

Then every x in $[a,b)$ has its corresponding $[x,x+h]$, so there exists a chain $[x_i,x_i+h_i]$ from a to b. If $B_{-\infty}$ and $B_{+\infty}$ denote the union of those $[x_i, x_i+h]$ with x_i in $e_{-\infty}$ and $e_{+\infty}$, respectively, then (27), (28), (27'), and (28') imply

$$v_1 + Mm(B_{-\infty}) + Mm(B_{+\infty}) \leq v + 2\epsilon(b - a),$$

in analogy with (30) of Theorem 5.8. Since $M \nearrow \infty$ as $\epsilon \searrow 0$ and since v remains bounded, then $m(B_{-\infty})$ and $m(B_{+\infty})$ tend to 0 with ϵ. It follows that $m(e_{-\infty}) = m(e_{+\infty}) = 0$ provided the sets $A_{-\infty}$ and $A_{+\infty}$ are chosen properly.

Theorem 5.8 also enabled Lebesgue to substantially improve the results in his thesis that deal with the concept of curve length. In the first place, simple modifications in the proof of Theorem 5.8 yield:

THEOREM 5.12.   *If C is a continuous curve defined by* $x = x(t)$, $y = y(t)$, $z = z(t)$ *for* $a \leq t \leq b$, *and if x, y, and z have bounded Dini derivatives, then these derivatives are summable and* x', y', z' *exist simultaneously almost everywhere. Furthermore, the length of C is given by*

$$* \int_a^b [(x')^2 + (y')^2 + (z')^2]^{1/2},$$

*where * indicates integration over all t such that x', y', and z' exist simultaneously.*

Theorem 5.12, although it at first appears limited in its scope because the Dini derivatives must be bounded, actually discredited the view of Scheeffer and others (3.4), since it shows that the concepts of curve length and integral are closely related: Suppose that C is any continuous, rectifiable curve with a parametric representation for which x, y, and z do not all remain constant on any interval. Then C can be represented using arc length s as parameter: $x = x_1(s)$, $y = y_1(s)$, $z = z_1(s)$. Then the Dini derivatives of x_1, y_1, and z_1 are bounded in absolute value by 1. By Theorem 5.12, x_1', y_1', and z_1' exist simultaneously almost everywhere and $s = * \int_a^b [(x_1')^2 + (y_1')^2 + (z_1')^2]^{1/2}$. "Rectifiable curves thus have tangents in general" [1904: 127]. And the classical integral formula for curve length is always valid if the parametric representation and the domain of integration are suitably chosen. Thus Lebesgue was able to reinstate the importance of the integral for the theory of curve rectification.

Lebesgue's results can easily be extended to rectifiable curves that are not necessarily continuous by the use of a result first given by Scheeffer: If $y = f(x)$ is rectifiable over $[a,b]$, then so is the continuous curve $y = \bar{f}(x) = f(x) - \varphi(x)$, where

$$\varphi(x) = [f(a+) - f(a)] + \sum_{a < t < x} [f(t+) - f(t-)] + [f(x) - f(x-)].$$

Furthermore, the length of $y = f(x)$ is equal to the length of $y = \bar{f}(x)$, which can be represented as a Lebesgue integral, plus the length of $y = \varphi(x)$, which is simply

$$|f(a+) - f(a)| + \sum_{a < x < b} |f(x+) - f(x-)| + |f(b) - f(b-)|.$$

For curves of the form $y = f(x)$, i.e., of the form $x = t$, $y = f(t)$—the class of rectifiable curves considered by Scheeffer—x and y can never be simultaneously constant, and the above reasoning applies. Lebesgue drew a further profound conclusion from this [1904: 127–128]. If, he observed, $x = x_1(s)$, $y = y_1(s)$ denotes the arc length representation of the curve $y = f(x)$, then $x_1'(s)$ and $y_1'(s)$ exist except on a set of values of s of measure zero. Since the mapping $x \to s$ "stretches" intervals, the corresponding set of x also has zero measure. Now if x is such that

$x_1'(s)$ and $y_1'(s)$ exist and $x_1'(s) \neq 0$, then $f'(x) = y_1'(s)/x_1'(s)$ exists. Furthermore, E, the set of x where $x_1'(s) = 0$, must have measure zero, since otherwise one of the Dini derivatives of f would be infinite on a set of positive measure,[13] contrary to Theorem 5.11. (It is at this point that Lebesgue indicated the proof of Theorem 5.11).

In summary, Lebesgue had proved the following:

> THEOREM 5.13.   *Any continuous function of bounded variation has a finite derivative almost everywhere.*

Thus Lebesgue's proof of Theorem 5.8, which suggested Theorem 5.11, and his results on the integral formula for curve length led to the discovery of Theorem 5.13. (Actually, Theorem 5.13 can be established independently of integration theory and without the assumption that f is continuous.) Theorem 5.13 in a sense substantiates the conviction of earlier mathematicians that functions are differentiable at most points. This theorem and Lebesgue's theorems on the integral formula for curve length and on term-by-term integration also in a sense provided a theoretical justification for what mathematicians during the eighteenth and early nineteenth centuries assumed without rigorous proof.

Harnack had called attention to the property of absolute continuity in connection with his extension of the Riemann integral to unbounded functions; he also attempted to relate absolute continuity to the extension proposed by Hölder (3.3). The term "absolute continuity" was not introduced until 1905 by G. Vitali and within the context of Lebesgue's theory of integration, but the property itself was frequently referred to during the 1890's when other mathematicians took up the study of the integral for unbounded functions.[14] Under the influence of Harnack's paper, these mathematicians all noted that if $\int_a^b |f|$ exists, then $F(x) = \int_a^x f$ is absolutely continuous. As Charles de la Vallée-Poussin expressed it, $\sum_{i=1}^n \int_{a_i}^{b_i} |f|$ converges to zero with $\sum_{i=1}^n (b_i - a_i)$. For Otto Stolz, absolute continuity represented one of the

---

13.   Since $x_1'(s)^2 + y_1'(s)^2 = 1$ for almost all s, those points of E for which $y_1'(s) = 0$ form a set of zero measure.

14.   De la Vallée-Poussin 1892; Jordan 1893–96: II, 49; Stolz 1898; Moore 1901a.

characteristic properties of the Riemann integral, a property also possessed by absolutely convergent Harnack integrals. E. H. Moore (1862–1932) of the University of Chicago considered absolute continuity a generalization of the property of uniform continuity.

Moore's work is particularly significant. In the first place, he called attention to some distinctions that Harnack had overlooked in his definition of the integral. Suppose that E is a subset of $[a,b]$ of content zero and that f is Riemann-integrable on every interval not meeting E. A union of intervals $U = \bigcup\limits_{i=1}^{n} I_i$ is said to cover E narrowly if each I_i meets E; if this is not necessarily the case, U is said to cover E broadly. This distinction becomes important when the integral is defined as the limit of $\int_a^b f_U$ as the content of U approaches 0, where $f_U(x) = 0$ if x is in U and $f_U(x) = f(x)$ otherwise. If the limit is taken over U's that cover E narrowly, the resulting integral is called the narrow integral; if the U's are permitted to cover E broadly, the resulting integral is called the broad integral. When a function f is broadly integrable, so is $|f|$.

Moore also pointed out that the characteristic distinction between these integrals is that broad integrals are absolutely continuous while narrow integrals need not be. To illustrate this point, Moore constructed the following example: Let $[a_n,b_n]$ denote an infinite sequence of intervals that are contained in $[a,b]$ and pairwise disjoint, such that a_n and b_n converge to b. Let $\sum\limits_{n=1}^{\infty} u_n$ denote any conditionally convergent series. Define f on $[a,b]$ in such a manner that f is continuous and of constant sign on each $[a_n,b_n]$, that $f(a_n) = f(b_n) = 0$, that $\int_{a_n}^{b_n} f = u_n$, and that $f(x) = 0$ for x not in the $[a_n,b_n]$. A particular example is the function f whose graph over $[a_n,b_n]$ forms an isosceles triangle with base $[a_n,b_n]$ and height $2u_n/(b_n-a_n)$. It is not difficult to see that f is continuous except at b, where it becomes unbounded, and that f is Harnack-integrable—i.e., f is narrowly integrable with respect to $E = \{b\}$. On the other hand, the integral of f is not absolutely continuous, since a finite subsequence u_n can be determined so that

$$\int_{a_{n_j}}^{b_{n_j}} f = \sum_j u_{n_j}$$

remains positive and large while $\sum\limits_j (b_{n_j} - a_{n_j})$ is made arbitrarily small.

This example also provided Moore with a counterexample to Harnack's proposition that a function which possesses a Hölder integral is also Harnack-integrable. Consider the closed set H consisting of the points a_n, b_n, and b. H is closed and is therefore the complement of open intervals with endpoints in H. A function g can easily be constructed so that: (i) g is unbounded in the neighborhood of a point if and only if the point belongs to H; (ii) the narrow integral of g with respect to $E = H$ exists; and (iii) $h = f + g$ also becomes unbounded at precisely the points of H. A theorem of Moore's shows that the narrow integral of f with respect to $E = H$ cannot exist; thus, h cannot be narrowly integrable with respect to $E = H$. On the other hand, h has an integral in Hölder's sense, since $F(x) = \int_a^x f + \int_a^x g$ satisfies the conditions for an indefinite Hölder integral of h, where $\int_a^x f$ is the narrow integral of f with respect to $E = \{b\}$ and $\int_a^x g$ is the narrow (or broad) integral of g with respect to $E = H$.

In summary, the property of absolute continuity came to be regarded as a characteristic property of an absolutely convergent integral. But no one actually attempted to show that any absolutely continuous function must be an absolutely convergent integral $\int_a^x f$, as that would involve establishing the existence of f. Du Bois-Reymond was the only mathematician who attempted that sort of proof when he tried to prove the existence of a Riemann-integrable function g such that $\underline{\int_a^x} f = \int_a^x g$, where f is an arbitrary bounded function. Du Bois-Reymond soon gave up this attempt, possibly because he saw that Riemann's definition was not general enough to permit the construction of g (2.2). Moore's example of a Hölder integral that is not a Harnack integral also vitiated Harnack's lone attempt to establish a formulation of the integral concept based on the existence of an indefinite integral that would be equivalent to the integral concept that proceeds from a limit definition of the definite integral.

Lebesgue was familiar with the work of de la Vallée-Poussin, Moore, and Harnack, and in the *Leçons* he was able to show that, in the context of his theory of integration, the concept of absolute continuity is precisely what is needed to reconcile the two approaches to the integral concept. First of all, in the case of bounded functions f defined on

$[a,b]$, Lebesgue defined f to be "summable" if there exists a continuous function F with bounded Dini derivatives such that $F'(x)=f(x)$ almost everywhere. Then $\int_a^b f$ is defined as $F(b)-F(a)$. That this definition is not ambiguous follows from Lebesgue's generalized version of a theorem due to Volterra and Scheeffer:

THEOREM 5.14. *If* $D^+F(x)=D^+G(x)$ *almost everywhere and* $D^+F(x)$ *and* $D^+G(x)$ *are bounded, then* $F-G$ *is constant.*

In an all-too-brief footnote at the end of the *Leçons* Lebesgue considered the more general case of possibly unbounded functions. He had observed in connection with Theorem 5.11 that the existence of continuous monotonic functions with dense intervals of invariability indicates that a continuous function of bounded variation need not be an integral. To be an integral, such a function, he observed, must also satisfy the condition of absolute continuity. Harnack, of course, realized that. But Lebesgue perceived that absolute continuity is actually sufficient, and he added: "If $f(x)$ is not subjected to the condition of being bounded nor $F(x)$ to that of having bounded derived numbers [Dini derivatives] but only to the preceding condition [i.e., absolute continuity], one has a definition of the integral equivalent to that developed in this chapter and applicable to all summable functions, bounded or not" [1904: 129n]. In other words:

THEOREM 5.15. *F is absolutely continuous on* $[a,b]$ *if and only if there exists a summable function f such that* $F(x)=\int_a^x f$ *for all x in* $[a,b]$.

Lebesgue did not bother to provide a proof of this proposition, and he probably did not forsee the usefulness of the notion of absolute continuity as fully as Giuseppe Vitali, who published the first proof and introduced the term "absolute continuity" (see 5.2). Nevertheless, the proof Lebesgue subsequently gave [1907b] is more interesting than Vitali's at this point because it is much simpler and, in its conception, is so much in line with the methods and results of the *Leçons* that it probably represents essentially what he had in mind when he announced Theorem 5.15.

It is not difficult to see that a Lebesgue integral is absolutely continuous; Lebesgue's accomplishment is that he established the converse. That is, suppose F is absolutely continuous. Then F is of bounded variation and hence, by Theorem 5.13, $F'(x)$ exists almost everywhere.

The obvious candidate for the f in Theorem 5.15 is the function f such that $f(x) = F'(x)$ at points x for which $F'(x)$ exists and $f(x) = 0$ otherwise. To show that $F(x) = \int_a^x f$, Lebesgue considered $g(x) = \int_a^x f - F(x)$. (It follows from the proof of Theorem 5.11 that f is actually summable.) By Theorem 5.10, $g'(x) = 0$ almost everywhere, and g is also absolutely continuous. Hence it remains to show that g must be constant. The method is essentially that employed by Lebesgue [1904: 78–79] to prove Theorem 5.14: Let ϵ be a positive number, and let E denote the set of points x such that $g'(x) \neq 0$ or $g'(x)$ fails to exist. E can be enclosed in A, a union of intervals of total length L, where L can be made arbitrarily small. For each t in $[a,x)$ there exists an interval $[t, t+h]$ such that if t is in E, then $[t, t+h]$ is contained in A, or $\left| g(t+h) - g(t) \right| \leq \epsilon h$ if $g'(t) = 0$. Then, by Theorem 5.9, there exists a chain of intervals $[t_i, t_i+h]$ from a to x. Now

$$\left| g(x) - g(a) \right| \leq \sum \left| g(t_i + h_i) - g(t_i) \right|$$

$$\leq {\sum}' \left| g(t_i + h) - g(t_i) \right| + \epsilon(b - a),$$

where $\sum'$ denotes summation over all t_i in A. Since g is absolutely continuous, the $\sum'$ term converges to zero with L, and the theorem is proved.

5.2   THE WORK OF VITALI AND YOUNG ON MEASURE AND INTEGRATION

The measure-theoretic ideas of Jordan and Borel set other mathematicians to thinking as well. In Italy, Giuseppe Vitali (1875–1932) developed a theory of measure similar to Lebesgue's. Vitali's interest in the measure of sets grew out of his discovery that the familiar necessary and sufficient condition for the Riemann integrability of a bounded function, (R_2), does not depend upon the nature of the discontinuities of the function [Vitali 1903]. At first glance such a dependency appears to exist, since the S_σ (the sets of points such that the oscillation of f, $\omega_f(x)$, is greater than or equal to σ) must have zero content. But, as Vitali knew, these sets are closed; so, if every closed subset of the set D of points of discontinuity of f has zero content, then f is integrable. Vitali discovered that the converse is also true: If f is integrable, then every closed subset of D has zero content. In other words, a bounded function is integrable if and only if D has the property that every closed subset of D has content zero. The integrability of f depends entirely upon the nature of the set D and not upon the behavior of f on the points of D.

In a subsequent paper [1904a], Vitali introduced a new concept, analogous to the outer content of Cantor and Jordan, so as to obtain a more succinct characterization of the set D corresponding to an integrable function f. He defined the minimal extension of a bounded set G to be the greatest lower bound of the sums of the lengths of all finite or countably infinite unions of nonoverlapping intervals that contain G. This is, of course, $m_e(G)$. Vitali observed that $c_i(G) \leq m_e(G) \leq c_e(G)$, so the minimal extension of a Jordan-measurable set is its content. After extending the Heine-Borel Theorem from an interval to an arbitrary closed, bounded set, Vitali proved that a bounded function f is integrable if and only if $m_e(D) = 0$.

Vitali was also acquainted with Borel's definition of measure and measurability, and in his next paper [1904b] he proceeded to develop his ideas into a theory of measure that satisfies the postulates prescribed by Borel and includes the definitions of Borel and Jordan as special cases. Starting from the notion of minimal extension, he observed that in general $m_e(G_1 \cup G_2) = m_e(G_1) + m_e(G_2) - Z$. Z, the "intertwinement" (*allaciamento*) of G_1 and G_2, is defined as the greatest lower bound of the total length of all unions C of nonoverlapping intervals formed by intersecting unions A and B of nonoverlapping intervals covering G_1 and G_2, respectively. Then a subset G of $[a,b]$ is defined to be measurable if the intertwinement of G and $[a,b] - G$ is zero. That is equivalent to Lebesgue's definition of measurability, and in fact Lebesgue had pointed out in his thesis that measurability could be characterized in this manner. Vitali also showed that properties (3), (6), (7), and (8) are satisfied by m_e restricted to measurable sets.

It was not until his paper was in press that Vitali learned of Lebesgue's announcement, in the *Comptes Rendus* for 1901, of his generalization of the notions of measure and integral. Clearly, when Vitali composed his paper he did not foresee, as had Lebesgue, the connection between his new theory of measure and a viable generalization of the integral.

Vitali quickly familiarized himself with the details of Lebesgue's theory and published a number of papers concerning it. In a paper of 1905 he introduced the concept and the term "absolute continuity" and proved Lebesgue's assertion that a function is absolutely continuous if and only if it is an integral [1904–05]. He also succeeded [1907] in completely resolving the question of term-by-term integration of nonuniformly bounded sequences as it had been discussed by Arzelà and

Osgood (see 4.4). Clearly, if $G(x) = \lim\limits_{n \to \infty} \int_a^x f_n$ is to equal $\int_a^x f$, then $G(x)$ must be absolutely continuous. By making some slight modifications in the proof of an earlier result of his on term-by-term integration [1905a], Vitali was able to show that the absolute continuity of G is also sufficient to ensure $G(x) = \int_a^x f$ when the X-points have zero measure. The following example, due to Vitali, illustrates the insufficiency of absolute continuity alone: Define f_n on $[0,1]$ so that $f_n(x) = n^2$ if x is in one of the intervals $[k/n, k/n + 1/n^3]$, for $k = 0, 1, \cdots, n-1$, and $f_n(x) = 0$ otherwise. Then $f_n(x)$ converges to 0 almost everywhere, but $\int_0^1 f_n = 1$. Every point is an X-point in this case.

Another mathematician, William Henry Young (1863–1942), actually went further than Vitali in duplicating Lebesgue's work and proposed a generalization of the integral similar to Lebesgue's. Young's career as a research mathematician began when, at the age of 34, he decided together with his wife, the mathematician Grace Chisholm Young, to devote himself to mathematical research; for this purpose they left England for Göttingen. During the period 1897–1908, while residing in Göttingen, Young produced an extraordinary number of papers in which can be seen his assimilation of the earlier developments in integration theory and the evolution of his own ideas on, among other things, the measure of sets and the concept of the integral. In discussing these papers we shall use the term "measure" where Young used "content," to have his terminology agree with Lebesgue's. Also it should be kept in mind that the term "open set" was used by Young to refer to any set that is not closed.

Young was familiar with Borel's *Leçons*, but did not entirely adopt Borel's theory of measure. For closed sets and sets formed by the countable union of nonoverlapping intervals, he accepted Borel's definitions [Young 1902–03a: 247–48]. Shortly thereafter he extended the definition of measure to sets formed by the union of overlapping intervals: From the given set he constructed a countable number of nonoverlapping intervals with the same interior points; he then defined the measure of the original set to be that of the constructed sets [Young 1902–03d]. The motivation for his subsequent extension of the definition of measure—which is based upon the introduction of the inner measure of an arbitrary set as the least upper bound of the measures of all closed subsets—probably was partly due to his discovery (similar to Vitali's) that the Riemann integrability of a bounded function f is not related to the values f assumes at D, the set of its points of discontinuity, since f is integrable if and only if the measure of every closed subset of D is zero [Young, 1904a]. A similar motivation may have come from his interest in Osgood's lemma (4.4) and in the question of its validity when G is not closed [Young 1902–03b, 1902–03c].

Osgood's paper also turned Young's interest to the problem of term-by-term integration, and Young became familiar with Arzelà's researches that had recently been republished and the fundamental lemma upon which they are based (4.4). Arzelà's proof of the lemma was not complete, and in the first part of the paper "Open Sets and the Theory of Content" [1904b], Young succeeded in securing the validity of the lemma and in generalizing it to:

> THEOREM 4.   *If an infinite collection of sets D, each a finite or infinite union of subintervals of $[a,b]$, has the property that the measure of each D exceeds a fixed number $g > 0$, then there exists a sequence $D_1, D_2, D_3, \cdots$ such that $\bigcap_{n=1}^{\infty} D_n$ contains closed subsets of measure greater than $g - \epsilon$ for any $\epsilon > 0$.*

The proof of this theorem is based on his Theorem 3: If $D_1, D_2, D_3, \cdots$ is a sequence of sets from the above collection such that D_{n+1} is a subset of D_n, then $\bigcap_{n=1}^{\infty} D_n$ contains closed subsets of measure greater than $g - \epsilon$ for any ϵ.

Young was able to show that Theorem 3 is still valid when the D_n are closed sets, a result he described as "very instructive and suggests at once the possibility of extending the theory of content to open [i.e., not closed] sets" [1904b: 23]; for Theorem 3, once it has been established for closed sets, can be easily extended to the following proposition.

> THEOREM 3'.   *If $G_1, G_2, G_3, \cdots$ is a sequence of sets such that $G_{n+1} \subset G_n$; and if the least upper bound of the measures of all closed subsets of G_n exceeds a fixed number $g > 0$ for every n, then $\bigcap_{n=1}^{\infty} G_n$ also contains closed subsets of measure greater than $g - \epsilon$ for any $\epsilon > 0$.*

Theorem 3' then yields a generalization of Theorem 4 in which the sets D of Theorem 4 are replaced by arbitrary sets G. Of course, the phrase "the measure of D_n" must also be changed to "the least upper bound of the measures of all closed subsets of G." In other words, "in the case of an open [i.e., not closed] set, the upper limit of the content of its closed components plays a most important role. In the lemmas and theorems relating to open [i.e., not closed] sets . . . this concept has to them precisely the relation that content itself has to closed sets" [1904b: 28]. Thus Young defined the inner measure of a set G as the least upper bound of the measures of its closed components.

Having extended the definition of measure (in the sense of an inner measure) to sets in general, Young considered the question of additivity. His approach is again significantly different from that of Lebesgue and Vitali for whom this question led directly to their definitions of measurability. Starting from the observation that if G_1 and G_2 are two disjoint sets such that one is closed, $m_i(G_1 \cup G_2) = m_i(G_1) + m_i(G_2)$, Young deduced a number of interesting results, which may be summarized as follows: Let $\mathcal{Y}_i$ denote the class of all sets G such that for all sets S that do not intersect G, $m_i(G \cup S) = m_i(G) + m_i(S)$. Then $\mathcal{Y}_i$ contains all closed sets, and $\mathcal{Y}_i$ and m_i satisfy (6), (7), and (8), the properties that Lebesgue had established for measurable sets.

The definition of the inner measure of a set is based upon the measure, or outer content, of a closed set. Applying the Heine-Borel theorem, Young observed that the measure of a closed set G can be defined as the greatest lower bound of the measures of all sets formed by enclosing each g in G in an interval. In this form the definition can be extended to any set G, and he defined the outer measure of G to be the above greatest lower bound. For closed sets $m_i(G) = m_e(G)$, and Young therefore proposed to call a set measurable whenever this is the case.

Next Young introduced the class $\mathcal{Y}_e$, defined analogously to $\mathcal{Y}_i$, and showed it also possesses properties (6), (7), and (8). The additive class $\mathcal{Y}$ is then defined to be $\mathcal{Y}_i \cap \mathcal{Y}_e$. Every measurable set is contained in $\mathcal{Y}_i$ and in $\mathcal{Y}_e$ and hence in $\mathcal{Y}$. The additive class $\mathcal{Y}$ particularly interested him, especially since "if there be other than measurable sets, it possesses distinct advantages over the class of measurable sets *in toto*. The fundamental property of additive sets embodied in the definition enables us to extend the theory of content to all sets of the additive class without scruple" [1904b: 49–50]. Young's interest in the theory of measure per se should be contrasted with Lebesgue's: For Lebesgue, the theory of measure was only a means to the end of generalizing the integral; consequently, the class of measurable sets was Lebesgue's primary interest.

In another paper [1904d], Young proposed to study the upper and lower integrals of a bounded function. These, he remarked, are considered mainly in order to state the necessary and sufficient condition for the existence of the Riemann integral. But he proposed to study them for their own sake, since it is "only a restricted class of functions that admit of integration in Riemann's sense . . ." [1904d: 52]. Young's main result was that the upper and lower integrals can be equated with certain ordinary integrals. The first step in establishing this is to show that corresponding to any bounded function f is an upper semi-continuous function, $\bar{f}$,[15] and a lower semi-continuous function, $\underline{f}$ such that

$$\overline{\int_a^b} f = \overline{\int_a^b} \bar{f} \quad \text{and} \quad \underline{\int_a^b} f = \underline{\int_a^b} \underline{f}.$$

Therefore the problem is reduced to that of semi-continuous functions. If $\bar{f}$ is upper semi-continuous, the set of x such that $\bar{f}(x) \geq k$ is closed for any real number k. Let $I(k)$ denote the measure of this set. Then I is an increasing function and hence Riemann-integrable. Similarly, if $\underline{f}$ is lower semi-continuous, the set of x such that $\underline{f}(x) \leq k$ is closed. If $J(k)$ denotes its measure, J is likewise Riemann-integrable. Young's result is that

(32)
$$\overline{\int_a^b} \bar{f} = m(b-a) + \int_m^M I(k)dk \quad \text{and}$$

$$\underline{\int_a^b} \underline{f} = M(b-a) - \int_m^M J(k)dk,$$

where $\underline{f}(x)$ and $\bar{f}(x)$ lie between m and M for all x in $[a,b]$.

What is especially interesting about this result is that these Riemann integrals are actually the Lebesgue integrals of $\bar{f}$ and $\underline{f}$. Consider, for example, $\bar{f}$: Let P be a partition of (m, M), $m = a_0 < a_1 < \cdots < a_n = M$. Then by definition $\int_m^M I(k)dk$ is the limit of the following sums as $\|P\|$ converges to 0, where e_i is the set of x such that $a_{i-1} \leq \bar{f}(x) < a_i$:

$$\sum_{i=1}^n I(a_{i-1})(a_i - a_{i-1})$$

$$= -a_0 I(a_0) + \sum_{i=1}^{n-1} a_i[I(a_{i-1}) - I(a_i)] + a_n I(a_{n-1})$$

$$= -(b-a)m + \sum_{i=1}^n a_i m(e_i).$$

15. Baire had introduced these notions in his thesis, which is referred to in 4.4. Young defined $\bar{f}(x)$ as follows: Let $M(d)$ denote the least upper bound of the numbers $f(t)$ for $|t-x| < d$. Then $\bar{f}(x) = \lim_{d \to 0} M(d)$; $\underline{f}(x)$ is defined analogously by use of greatest lower bounds. De la Vallée-Poussin had introduced the two-variable analogue of $\underline{f}$ to treat Fubini's Theorem (5.3).

Hence, $(b-a)m+\displaystyle\int_m^M I(k)dk$ is the limit of $\displaystyle\sum_{i=1}^n a_i m(e_i)$, i.e., is the Lebesgue integral of $\bar{f}$. The expressions $(b-a)m+\displaystyle\int_m^M I(k)dk$ and $(b-a)M-\displaystyle\int_m^M J(k)dk$, of course, remain meaningful as long as the sets used to define I and J are measurable, that is, as long as f is measurable. In that case, the expressions will coincide and therefore provide a generalization of the integral. But Young did not pursue this line of thought. Instead he observed that these expressions remain meaningful when $[a,b]$ is replaced by an arbitrary closed set S, and he thus proposed to define $\displaystyle\overline{\int_S} f$ and $\displaystyle\underline{\int_S} f$ by equating them with these expressions, $I(k)$ corresponding to $\bar{f}$ and $J(k)$ to $\underline{f}$.

It was by a different course of reasoning that Young was led to his generalization of the integral. His interest in the upper and lower integrals led him to study "whether it is admissible to adopt a more general mode of division of the fundamental segment than that used by Riemann, Darboux and other writers, when forming summations (upper, lower summations), defining as limit the integral (upper, lower integral) of a function over the fundamental segment" [1904c: 445]. This problem was probably motivated by his interest in extending the definition of the integral to sets more general than intervals, as well as by the form of Jordan's definition of the Riemann integral.

The above quotation is from an abstract of a lengthy paper published by Young [1905] in the *Philosphical Transactions*. After a few digressions, Young [1905] began by introducing the following tentative generalization of the definition of upper and lower integrals: Let e_1, e_2, e_3, $\cdots$ denote a partition of $[a,b]$ into pairwise disjoint, measurable sets. Then $(Y)\displaystyle\overline{\int_a^b} f$ is defined to be the greatest lower bound of the sums $\sum M_i m(e_i)$, and $(Y)\displaystyle\underline{\int_a^b} f$ the least upper bound of the sums $\sum m_i m(e_i)$, where m_i and M_i denote the greatest lower bound and the least upper bound, respectively, of f on e_i. (The notation (Y) is not Young's.) The terms upper and lower integral for these numbers, he added, are not justified unless

$$(33) \qquad\qquad (Y)\underline{\int_a^b} f \le (Y)\overline{\int_a^b} f$$

and unless these numbers coincide with the numbers obtained in the customary definition of the upper and lower integral. Although (33) is true, because, as he showed,

$$(34) \qquad \underline{\int_a^b} f \leq (Y) \underline{\int_a^b} f \leq (Y) \overline{\int_a^b} f \leq \overline{\int_a^b} f,$$

the second condition does not generally hold: for if f is Dirichlet's function—i.e., $f(x) = \chi_Q(x)$, where Q denotes the set of rational numbers—then $(Y) \underline{\int_0^1} f = (Y) \overline{\int_0^1} f = 0$, whereas $\underline{\int_0^1} f = 0$ and $\overline{\int_0^1} f = 1$.

The inequalities in (34) indicate that Young's definitions could be used to generalize the usual notion of integrability; Dirichlet's function illustrates the broader scope thereby obtained for the resulting concept of the integral. But Young was not ready to jump to these conclusions yet. Instead he took a course similar to that taken in his paper on upper and lower integrals. Thus he showed that

$$\underline{\int_a^b} f = (Y) \underline{\int_a^b} f; \qquad \overline{\int_a^b} f = (Y) \overline{\int_a^b} f.$$

These relationships then enabled him to extend the ordinary upper and lower integrals to any measurable set—not just to closed sets as in his earlier paper—by defining $\underline{\int_S} f$ to be $(Y) \underline{\int_S} f$ and $\overline{\int_S} f$ to be $(Y) \overline{\int_S} f$, since his integrals are defined for any measurable set. But Young discovered that this extension of the customary definitions is not without drawbacks: for if R and Q denote the sets of irrational and rational numbers, respectively, in $[0,1]$ and if $f = \chi_Q$ is Dirichlet's function, then f is integrable over both R and Q in the sense that $\underline{\int_R} f = \overline{\int_R} f = 0$ and $\underline{\int_Q} f = \overline{\int_Q} f = 0$; but f is not integrable over the union of R and Q. In short, Young's new ideas were really not compatible with the older ones.

And so after these detours, Young returned to his definitions of $(Y) \underline{\int_S} f$ and $(Y) \overline{\int_S} f$, "which we saw did not agree with the usual definitions. On the other hand the definitions we have since constructed seem more artificial than these. It suggests itself, therefore, that the most logical plan is to throw overboard the Riemann and Darboux definitions altogether . . . " [1905:243]. The numbers $(Y) \underline{\int_S} f$ and $(Y) \overline{\int_S} f$ were then termed the inner and outer measures of the integral. When they are equal, their common value is taken to be the integral of f.

By the time he wrote up these ideas for publication, Young had learned of Lebesgue's work. He concluded his paper by showing, by means of identities similar to (29) (but involving $(Y) \overline{\int_S} \bar{f}$ and $(Y) \underline{\int_S} f$) that when f is bounded and measurable, Lebesgue's definition and his own coincide.

There are some notable similarities and differences between Young's paper and Lebesgue's thesis. Both were inspired by the ideas of Jordan and Borel on the measure of sets, and both present generalizations of the integral that are essentially the same. But Young's paper led up to, and climaxed in, his generalized definition of the integral. In Lebesgue's thesis, the generalization of the integral was only the starting point, the basis upon which a substantial theory was then constructed. In Young's paper, the new definition of the integral was obtained by straightforward generalization from Jordan's version of Riemann's definition. A subtler generalization is presented in Lebesgue's thesis and is based upon partitions of the range of the integrand. We have already indicated the important role these partitions played in Lebesgue's successful development of the theory of the generalized integral.

5.3   FUBINI'S THEOREM

We adopted the convention of using "Fubini's Theorem" as a generic title to refer to any theorem positing the identity of a double integral with iterated integrals. When this theorem was extended to the class of Riemann-integrable functions (4.1), certain modifications in its formulation became necessary to account for the fact that the integrability of $f(x,y)$ as a function of two variables does not entail the integrability of the functions $x \to f(x,y)$ and $y \to f(x,y)$. Further difficulties were encountered when attempts were made to extend Fubini's theorem to unbounded functions. As a result, its formulation became complicated and in fact scarcely recognizable until Fubini, building upon Lebesgue's treatment of the problem, was able to reestablish the theorem in the context of Lebesgue's theory of integration and to restore in essence the simplicity of its classical formulation.

That a satisfactory formulation of Fubini's Theorem for unbounded functions would be a formidable problem was first indicated by de la

Vallée-Poussin [1892] when he extended Riemann's definition of the integral to unbounded functions. For a function of one variable, the definition is extended in the following manner: Suppose that f is defined on $[a,b]$ and that U, the set of x such that $\omega_f(x) = \infty$, has zero content. If x is in U, then $f(x)$ can be infinite or f can be unbounded in the neighborhood of x. Suppose further that f is Riemann-integrable on any subinterval not meeting U. Let $N_1(n)$ and $N_2(n)$ denote two sequences of positive numbers such that $\lim_{n \to \infty} N_1(n) = \lim_{n \to \infty} N_2(n) = \infty$; let $f_n = -N_1(n)$ if $f(x) < -N_1(n)$; $f_n(x) = f(x)$ if $-N_1(n) \le f(x) \le N_2(n)$; and $f_n(x) = N_2(n)$ if $f(x) > N_2(n)$. Then f is defined to be integrable over $[a,b]$ if the integrals $\int_a^b f_n$ converge to a finite limit that is independent of the choice of the sequences $N_1(n)$ and $N_2(n)$. When this limit exists, it is by definition $\int_a^b f$. De la Vallée-Poussin's integral is an absolutely convergent integral in the sense that f is integrable if and only if $|f|$ is integrable.

For simplicity, in discussing Fubini's Theorem, de la Vallée-Poussin assumed that T is a subset of the plane such that lines parallel to the coordinate axes meet T in an interval. He also assumed that if $[x_1, x_2]$ denotes the intersection of T with the line $Y = y$, then

$$I(y) = \int_{x_1}^{x_2} f(x,y)dx$$

exists as a finite number or is $+\infty$ or $-\infty$ for each y. It follows from Jordan's version of Fubini's Theorem that when f is bounded and $I(y)$ exists, then $I(y)$ is integrable and $\int_c^d I(y) = \int_T f(x,y)dT$. But when f is unbounded, $I(y)$ need not be integrable, as de la Vallée-Poussin discovered.

This difficulty is related to what he referred to as the "regular convergence" of $I(y)$. Specifically, $I(y)$ is said to converge regularly if for every positive ϵ there exists a set $E(\epsilon)$ of zero content and an integer n such that

$$\left| \int_{x_1}^{x_2} |f(x,y)| \, dx - \int_{x_1}^{x_2} |f_n(x,y)| \, dx \right| < \epsilon$$

for all y not in $E(\epsilon)$. The significance of regular convergence is expressed in the following proposition:

Let $f(x,y) \geq 0$ be integrable over T. Then $I(y)$ is integrable and $\int_c^d I(y)dy = \int_T f(x,y)dT$ if and only if $I(y)$ is regularly convergent.

A simple example illustrating the sort of difficulties that arise with de la Vallée-Poussin's definition was given by Otto Stolz in the third volume of his *Grundzüge der Differential- und Integralrechnung* [1893–99: III, 139–40], which is entirely devoted to the integration of functions of several variables and relies heavily upon de la Vallée-Poussin's paper. Let T be the unit square $0 \leq x \leq 1$, $0 \leq y \leq 1$; consider $f(x,y) = t(x)g(y)$, where $t(x) = 1/2^n$ if $x = (2m+1)/2^n$ and $t(x) = 0$ otherwise, and where g is a nonnegative function continuous on $(0,1)$ but unbounded in the neighborhood of 0 or 1 in such a manner that $\int_0^1 g = \infty$.

Then f is integrable over T and its integral is 0, but $I(x) = \int_0^1 f(x,y)dy$ is not integrable, since $I(x) = +\infty$ if $x = (2m+1)/2^n$ and $I(x) = 0$ otherwise. $I(x)$ is akin to Dirichlet's example of a nonintegrable function.

Actually de la Vallée-Poussin was able to establish only the "only if" part of the above proposition and only for functions satisfying an additional condition. Later E. W. Hobson (1856–1933) succeeded in removing that condition at least for nonnegative functions [1906]. Hobson did not assume the existence of $I(y)$ either, so the definition of regular convergence had to be generalized accordingly. Thus, for nonnegative functions he was able to establish the necessity and sufficiency of the condition of generalized regular convergence for the validity of Fubini's Theorem. Hobson's version of Fubini's Theorem is the following:

If $\overline{I}_n(y) = \int_{x_1}^{x_2} f_n(x,y)dx$, $\underline{I}_n(y) = \int_{\underline{x_1}}^{x_2} f_n(x,y)dx$, $\overline{I}(y) = \limsup_{n \to \infty} \overline{I}_n(y)$, and $\underline{I}(y) = \liminf_{n \to \infty} \underline{I}_n(y)$, then $\overline{I}(y)$ and $\underline{I}(y)$ are integrable, and $$\int_c^{\overline{d}} \overline{I}(y)dy = \int_{\underline{c}}^d \underline{I}(y)dy = \int_T f(x,y)dT.$$

An equally complicated version of Fubini's Theorem had been obtained by de la Vallée-Poussin [1899]. Here he introduced the lower semi-continuous function $mf(x,y)$, which is constructed from $f(x,y)$ in the following manner: let $m(d)$ denote the greatest lower bound of the numbers $f(s,t)$ such that $[(s-x)^2 + (t-y)^2]^{1/2}$ is less than d. Then $mf(x,y) = \lim_{d \to 0} m(d)$. Using $mf(x,y)$ he was able to prove the following formulation of Fubini's Theorem:

If $f(x,y) \geq 0$ is integrable, then

$$\int_T f(x,y)dT \ = \ \underline{\int_a^b} dx \left[\underline{\int_c^d} mf(x,y)dy \right] = \underline{\int_c^d} dy \left[\underline{\int_a^b} mf(x,y)dx \right].$$

For functions of variable sign, the theorem must be applied separately to $f^+(x,y) = \max[f(x,y),0]$ and $f^-(x,y) = \max[-f(x,y),0]$, thus compounding the complexity of the formulation.

In his thesis, Lebesgue had also treated the relationship between double and repeated integrals and had encountered the same type of difficulty as his predecessors: If $f(x,y)$ is summable on a measurable set E, the functions $x \to f(x,y)$ and $y \to f(x,y)$ need not be measurable [1902a: 276ff]. Like them he proceeded to introduce upper and lower integrals to deal with these nonmeasurable functions. Suppose $f(x)$ is defined and bounded on a measurable set E. Let $\mathcal{L}$ denote the class of all bounded, summable functions φ such that $\varphi(x) < f(x)$ for all x. The least upper bound L of all numbers $\int_E \varphi$ for φ in $\mathcal{L}$ is by definition the lower integral of f: $\underline{\int_E} f = L$. The upper integral $\overline{\int_E} f$ is defined similarly. (It follows easily from Lebesgue's remarks about these definitions that $\underline{\int_E} f \leq \overline{\int_E} f$ and that f is summable on E if and only if $\underline{\int_E} f = \overline{\int_E} f$.)

Lebesgue proved the following result:

THEOREM 5.16. *If $f(x,y)$ is bounded and measurable on the rectangle R defined by $a \leq x \leq b$, $c \leq y \leq d$, then*

$$\int_R f(x,y)dR \ = \ \underline{\int_c^d} dy \left[\underline{\int_a^b} f(x,y)dx \right] = \underline{\int_a^b} dx \left[\underline{\int_c^d} f(x,y)dy \right]$$

$$= \ \overline{\int_c^d} dy \left[\overline{\int_a^b} f(x,y)dx \right] = \overline{\int_a^b} dx \left[\overline{\int_c^d} f(x,y)dy \right].$$

As Lebesgue noted, Theorem 5.16 can be extended immediately to a function g defined on a measurable set E contained in R by applying Theorem 5.16 to the function $f(x,y)$ that is equal to $g(x,y)$ for (x,y) in E and to 0 for (x,y) in $R-E$. In view of this fact, we shall restrict the following discussion of Fubini's Theorem to double integrals over rectangles.

It should be noted that, in Theorem 5.16, both iterated integrals must be lower (or upper) integrals, in contrast to the version of Fubini's Theorem for Riemann integrals. But Lebesgue made an observation that suggested to Fubini the possibility of improving Theorem 5.16: If the function f of Theorem 5.16 is Borel-measurable—that is, if for every real number a the set of points (x,y) such that $f(x,y) > a$ is Borel-measurable—then the functions $x{\rightarrow}f(x,y)$ and $y{\rightarrow}f(x,y)$ are also Borel-measurable. (Here Lebesgue made use of the fact that the intersection of a planar Borel-measurable set and a line parallel to one of the coordinate axes is a linear Borel-measurable set.) Hence for Borel-measurable functions, the upper and lower integrals in Theorem 5.16 can be replaced by ordinary Lebesgue integrals.

In the paper [1906] in which Hobson treated de la Vallée-Poussin's notion of regular convergence, he also put the difficulties encountered by de la Vallée-Poussin in a new light. If (in the notation of Hobson's version of Fubini's Theorem) $\underline{I}_n = \overline{I}_n$ and $\underline{I} = \overline{I}$, then the generalized definition of regular convergence becomes Arzelà's necessary and sufficient condition that $I(y) = \lim_{n \to \infty} I_n(y)$ satisfy Riemann's integrability condition (R$_2$) (without the additional condition that $I(y)$ be finite-valued or bounded). In other words, it is precisely the fact that this integrability condition is not always preserved in the limit that is behind the difficulties encountered by de la Vallée-Poussin.

In a subsequent paper [1907], Hobson went further and stressed that most of the specific exceptions to Fubini's Theorem that had been constructed cease to be exceptions when Lebesgue integrals are introduced. Thomae [1878] had introduced the function $f(x,y)$ defined on the unit square $R(0 \le x \le 1, \, 0 \le y \le 1)$ as follows: $f(x,y) = 2y$ if x is irrational; $f(x,y) = 1$ if x is rational. Then $\displaystyle\int_0^1 dx \left[\int_0^1 f(x,y)dy \right] = 1$ by Riemann's definition. But f is discontinuous at every point in R except those on the line $y = 1/2$, and $x{\rightarrow}f(x,y)$ is totally discontinuous for $y \ne 1/2$. Thus the integrals $\displaystyle\int_R f(x,y)dR$ and $\displaystyle\int_0^1 dy \left[\int_0^1 f(x,y)dx \right]$ do not exist in Riemann's sense. Hobson pointed out that both of these integrals exist in Lebesgue's sense, however, and both have the value 1.

Next Hobson considered an example due to A. Pringsheim [1900: 48–51]: Let R denote the unit square. Consider the set of points (x,y) in R such that x and y have finite decimal expansions of the same length; that is, for some integer k, $x = .n_1 n_2 \cdots n_k$, $y = .m_1 m_2 \cdots m_k$, and $n_k m_k \ne 0$. This set, call it E, is a subset of the set of points in R with rational coordinates. It is not difficult to see that E is dense in R but that lines parallel to the coordinate axes meet E in at most a finite number of points. Let $c \ne c'$, and define f on R by: $f(x,y) = c'$ if (x,y) is in E; $f(x,y) = c$ if (x,y) is in $R - E$. Because E is dense, $\int_R f(x,y)dR$ does not exist in Riemann's sense, but the functions $x \to f(x,y)$ and $y \to f(x,y)$ are Riemann-integrable, since they take the value c except possibly for a finite number of points. Hence $\int_0^1 dy \left[\int_0^1 f(x,y)dx \right] =$ $\int_0^1 dx \left[\int_0^1 f(x,y)dy \right] = c$. And once again, Hobson noted, $\int_R f(x,y)dR$ exists as a Lebesgue integral and has the value c, since $f(x,y) = c'$ on a countable set, that is, on a set of measure zero.

Turning to examples involving unbounded functions, Hobson considered Stolz's function $f(x,y) = t(x)g(y)$ discussed earlier. The Riemann integral of f on the unit square R is zero; $\int_0^1 f(x,y)dx = g(y) \int_0^1 t(x)dx = 0$, and so $\int_0^1 dy \left[\int_0^1 f(x,y)dx \right] = 0$. But $\int_0^1 dx \left[\int_0^1 f(x,y)dy \right]$ does not exist in accordance with de la Vallée-Poussin's definition because $\int_0^1 f(x,y)dy = +\infty$ for x in the dense set of points of the form $(2m+1)/2^n$. This set, however, has Lebesgue measure zero. "Hence, since the Lebesgue integral is independent of the functional values at a set of points of zero measure, . . . $\left[\int_0^1 f(x,y)dy \right]$ is integrable with respect to x in the interval $(0,1)$ and has the value zero" [1907: 331].

This seemingly innocent remark by Hobson requires some clarification. Suppose g is defined and summable on $[0,1]$. Let $E \subset [0,1]$ have measure zero. If the values g takes on E are changed so that a new function g^* is defined on $[0,1]$, then, from Lebesgue's analytical definition of the integral, g^* is summable and $\int_0^1 g^* = \int_0^1 g$. The following extension of Lebesgue's definition is thus suggested: If g is defined on $[0,1]$ except on a set E of zero measure and if g is summable in Lebesgue's sense on $[0,1] - E$, then define $\int_0^1 g$ to be the Lebesgue integral $\int_{E^c} g$, where $E^c = [0,1] - E$. Lebesgue himself permitted such an extension only when E is finite. Thus, for example, Lebesgue expressed the arc length formula with the notation $s = \int_{\mathcal{E}} [(x')^2 + (y')^2 + (z')^2]^{1/2}$ in Theorem 5.12, where $\mathcal{E} = [a,b] - E$ and $m(E) = 0$, instead of extending his definition of the integral. Hobson, however, tacitly accepted the extended version of the Lebesgue integral and was therefore able to say

for Stolz's function that $\displaystyle\int_R f(x,y)dR = \int_0^1 dx\left[\int_0^1 f(x,y)dy\right].$

Hobson failed to realize, however, that his observations about the validity of Fubini's Theorem for Stolz's function are true in general. This discovery was made by two Italian mathematicians, Beppo Levi and Guido Fubini (1879–1943). In a footnote to a paper [1906e] on Dirichlet's Principle, Levi commented in passing that if f is bounded and summable on the unit square, "it would certainly be possible to conclude that the linear integrals exist on every $y = $ const., except at most for lines of an aggregate of measure zero. . . . Making abstraction from the aggregate of these lines (which is possible because it has superficial measure zero), the Lebesgue integral over the area can then always be obtained by two successive integrations." [1906e: 322n] He did not, however, provide any proof of his remark or consider the case of unbounded functions. Independently of Levi, Fubini [1907] discovered the following result:

THEOREM 5.17. *If $f(x,y)$ is summable on the rectangle R defined by $a \le x \le b$, $c \le y \le d$, then the functions $x \rightarrow f(x,y)$ and $y \rightarrow f(x,y)$ are summable for almost all values of y and x, respectively. Furthermore, the functions $y \rightarrow \int_a^b f(x,y)dx$ and $x \rightarrow \int_c^d f(x,y)dy$ are summable [in the extended sense] and*

$$\int_R f(x,y)dR = \int_c^d dy\left[\int_a^b f(x,y)dx\right] = \int_a^b dx\left[\int_c^d f(x,y)dy\right].$$

Fubini was led to Theorem 5.17 for bounded functions by Hobson's observation—made independently by Fubini—concerning the negligibility of sets of zero measure and by Lebesgue's comments on Theorem 5.16 for Borel-measurable functions. With these functions in mind, Fubini defined a function $f(x,y)$ to be linearly measurable if the functions $x \rightarrow f(x,y)$ and $y \rightarrow f(x,y)$ are measurable. Lebesgue's Theorem 5.16, of course, implies Theorem 5.17 for bounded functions that are measurable and also linearly measurable. Moreover, Lebesgue's Theorem 5.1 implies that any measurable set E can be expressed as $E = B \cup N$, where B is a Borel set and $m(N) = 0$. Fubini's idea was to use this property of measurable sets to establish the existence of a bounded, measurable, linearly measurable function φ, equal to the given function f almost everywhere, such that for almost every y the functions $x \rightarrow f(x,y)$ and $x \rightarrow \varphi(x,y)$ are also equal almost everywhere.

The validity of Theorem 5.17 for φ then implies its validity for f (using Lebesgue integrals in the extended sense).[16]

To extend Theorem 5.17 to unbounded functions, Fubini applied a theorem proved by Levi [1906d]. Levi discovered that if f_n is a non-decreasing sequence of summable functions defined on a measurable set E such that $\lim\limits_{n\to\infty} \int_E f_n$ is finite, then $f(x) = \lim\limits_{n\to\infty} f_n(x)$ is finite almost everywhere and is summable with $\int_E f = \lim\limits_{n\to\infty} \int_E f_n$. Fubini applied that result as follows: Let $K_n \nearrow \infty$ and assume without loss of generality that f is nonnegative. If $f_n(x,y)$ is defined to equal $f(x,y)$ when (x,y) is such that $f(x,y) \leq K_n$ and to be 0 otherwise, then $\lim\limits_{n\to\infty} \int_R f_n = \int_R f$. Also as f_n satisfies the conditions of Theorem 5.17 in the cases already established,

$$v_n(x) = \int_c^d f_n(x,y)\,dy$$

exists almost everywhere. The sequence v_n satisfies the conditions of Levi's theorem, since it is increasing and

$$\lim_{n\to\infty} \int_a^b v_n(x)\,dx = \lim_{n\to\infty} \int_a^b dx \left[\int_c^b f_n(x,y)\,dy \right] = \lim_{n\to\infty} \int_R f_n = \int_R f.$$

Thus $v(x) = \lim\limits_{n\to\infty} v_n(x)$ is finite almost everywhere and summable, and

$$\int_R f = \lim_{n\to\infty} \int_a^b v_n = \int_a^b v = \int_a^b dx \left[\lim_{n\to\infty} \int_c^d f_n(x,y)\,dy \right]$$
$$= \int_a^b dx \left[\int_c^d f(x,y)\,dy \right].$$

The appearance of Fubini's paper marked a real triumph for Lebesgue's ideas. As Fubini said, the Lebesgue integral "is now necessary in this type of study" [1907: 608].

De la Vallée-Poussin later [1910: 769–770] made the interesting observation that for bounded functions Theorem 5.17 follows quite simply from Lebesgue's Theorem 5.16. Theorem 5.16 implies that

---

16. Fubini's proof of Theorem 5.17 was defective: the function φ defined by him [1907: 610] need not be linearly measurable. Entirely different proofs were supplied by Hobson [1910] and de la Vallée-Poussin [1910].

$$0 \leq \overline{\int}_a^b dx \left[\overline{\int}_c^d f(x,y)dy \right] - \int_a^b dx \left[\overline{\int}_c^d f(x,y)dy \right]$$

$$\leq \overline{\int}_a^b dx \left[\overline{\int}_c^d f(x,y)dy \right] - \int_a^b dx \left[\underline{\int}_c^d f(x,y)dy \right] = 0.$$

Hence the upper and lower integrals of $\overline{\int}_c^d f(x,y)dy$ coincide and

$\overline{\int}_c^d f(x,y)dy$ is summable as a function of x. Similarly $\underline{\int}_c^d f(x,y)dy$ is

summable. Thus Theorem 5.16 states that

$$0 = \int_a^b dx \left[\overline{\int}_c^d f(x,y)dy \right] - \int_a^b dx \left[\underline{\int}_c^d f(x,y)dy \right]$$

$$= \int_a^b dx \left[\overline{\int}_c^d f(x,y)dy - \underline{\int}_c^d f(x,y)dy \right].$$

Thus, the nonnegative function $\overline{\int}_c^d f(x,y)dy - \underline{\int}_c^d f(x,y)dy$ is zero almost

everywhere, i.e., $y{\to}f(x,y)$ is summable for almost all x. Consequently,

$\int_a^b dx \left[\underline{\int}_c^d f(x,y)dy \right] = \int_a^b dx \left[\int_c^d f(x,y)dy \right]$, where the right-hand

side is a Lebesgue integral in the extended sense: Theorem 5.16 is thereby
transformed into Theorem 5.17. Of course, Lebesgue himself did not
consider such an extension of his definition.

6

Pioneering Applications of the Lebesgue Integral

During the period 1902–07 a number of important papers appeared, by Lebesgue, Fatou, Riesz, and Fischer, demonstrating the merits of the Lebesgue integral in applications to other branches of mathematics. Through his researches on trigonometric series, Lebesgue established in particular the utility of his theorem on term-by-term

integration and of his theorem that $\left(\int_a^x f \right) = f(x)$ almost everywhere.

Fatou's study of trigonometric series and analytic functions follows in the footsteps of Lebesgue's work. But Fatou was the first to demonstrate the usefulness of Lebesgue's theorems on the differentiation of continuous functions of bounded variation. These theorems were later applied by Riesz and Fischer to obtain, independently of each other, the Riesz-Fischer Theorem. In fact, Riesz's proof of this theorem is a simple modification of one of Fatou's proofs. The Riesz-Fischer Theorem and Riesz's subsequent work on L^p spaces (1910) helped secure a permanent place for Lebesgue's theory in the theory of integral equations and function spaces.

Riemann's theory of integration, both in its origins and in its subsequent development, was closely related to investigations in the theory of trigonometric series. Consequently it is not surprising that Lebesgue's first application of his integral was in this area. "In concerning myself with trigonometric series," Lebesgue wrote in his first detailed paper on the subject, "my principal objective has been to demonstrate the utility that the notion of the integral that I introduced in my thesis could have in the study of discontinuous functions of a real variable" [1903a: 453]. And it is in the theory of trigonometric series that he demonstrated this utility most successfully.

The first problem that Lebesgue [1902b][1] attacked was that of extending du Bois-Reymond's theorem on the nature of the coefficients a_n and b_n in a trigonometric series representation of a bounded function f, viz.,

$$(1) \qquad f(x) = \tfrac{1}{2}a_0 + \sum_{n=1}^{\infty} (a_n \cos nx + b_n \sin nx).$$

Fourier had "proved" that a_n and b_n must be the Fourier coefficients of f by integrating the right-hand side of (1) term by term; du Bois-Reymond was able to justify Fourier's conclusion when f is Riemann-integrable by employing a rather complicated argument that avoids assuming the permissibility of term-by-term integration. (See 1.1 and 2.1.)

In effect, what Lebesgue discovered was that Fourier's assumption and his method for proving it are basically sound within the context of Lebesgue's definition of the integral. In the first place, any function given by (1) is measurable, since it is the limit of a sequence of continuous functions. Since f is bounded, it is summable. The proof that a_n and b_n are the Fourier coefficients of f is based upon the following lemma:

LEMMA 6.1. *If $F(x)$ denotes Riemann's function (defined by (4) of 1.4) corresponding to the series in (1), then*

$$\frac{\Delta^2 F(x)}{\alpha^2} = \frac{F(x + \alpha) + F(x - \alpha) - 2F(x)}{\alpha^2}$$

lies between the greatest lower bound and least upper bound of $f(t)$ for t in $[x-\alpha,\ x+\alpha]$.

Thus, if f is bounded, $\Delta^2 F(x)/\alpha^2$ is uniformly bounded in x and α. This property can be used in two ways with Lebesgue's Theorem 5.3 on term-by-term integration to prove that a_n and b_n are the Fourier coefficients.

The first way, which is the way that Lebesgue originally obtained the theorem, is based upon consideration of the function

$$(2) \qquad f(r,u) = \tfrac{1}{2}\, a_0 + \sum_{n=1}^{\infty} (a_n \cos nu + b_n \sin nu)r^n,$$

---

1.   The proof is given in more detail in [Lebesgue 1903a].

where $0 \leq u \leq 2\pi$ and $0 \leq r \leq 1$. This function, as was well known, is harmonic for r less than 1, and, by a theorem due to Abel, $f(u) = \lim_{r \to 1} f(r,u)$.

For any r less than 1 the series in (2) converges uniformly; term-by-term integration yields

(3)
$$\frac{1}{\pi} \int_0^{2\pi} f(r,u) \cos nu \; du = a_n r^n;$$

$$\frac{1}{\pi} \int_0^{2\pi} f(r,u) \sin nu \; du = b_n r^n.$$

Using Lemma 6.1 and the properties of harmonic functions, Lebesgue showed that $f(r,u)$ is uniformly bounded in r. Thus, letting $r \nearrow 1$ in (3) and applying Theorem 5.3, Lebesgue obtained

$$\frac{1}{\pi} \int_0^{2\pi} f(u) \cos nu \; du = a_n;$$

$$\frac{1}{\pi} \int_0^{2\pi} f(u) \sin nu \; du = b_n.$$

The second method of proof is that used by du Bois-Reymond and consists in showing that Riemann's function F can be expressed in the form

(4)
$$F(u) = \int_0^u dv \left[\int_0^v f(t) dt \right] + Au + B.$$

In view of Lemma 6.1, $\Delta^2 F(u)/\alpha^2$ is uniformly bounded in α. Also, Riemann had proved that $f(u) = \lim_{\alpha \to 0} \Delta^2 F(u)/\alpha^2$ (formula (5) of 1.4). Theorem 5.3 thus applies and yields

$$\int_0^u dv \left[\int_0^v f(t) dt \right] = \lim_{\alpha \to 0} \int_0^u dv \left[\int_0^v \Delta^2 F(t)/\alpha^2 dt \right],$$

and straightforward computation of the right-hand side yields an expression of the form $F(u) + Au + B$. Hence, (4) is proved. The elegant simplicity of Lebesgue's proofs must be compared with the complexity of du Bois-Reymond's proof, which establishes a less general result.

Lebesgue [1905a; 1905b] applied his theory of integration to the study of the sequence of arithmetic means $\sigma_n(x)$ corresponding to the partial sums $S_n(x)$ of the Fourier series of a function f:

$$\sigma_n(x) = \frac{S_0(x) + S_1(x) + \cdots + S_{n-1}(x)}{n}.$$

The arithmetic means had been introduced by Cesàro in 1890 and had been applied to the theory of Fourier series by the Hungarian mathe-

matician Leopold Fejér (1880–1959). Fejér [1900; 1904] showed that if
f is Riemann-integrable, then

$$\sigma_n(x) - f(x) - \frac{1}{n\pi} \int_0^{\pi/2} g_x(t) \frac{\sin^2 nt}{\sin t} \, dt,$$

where $g_x(t) = f(x+2t) + f(x-2t) - 2f(x)$. Using the positivity of
$\dfrac{\sin^2 nt}{\sin t}$, he succeeded in showing that $\lim\limits_{n\to\infty} \sigma_n(x) = f(x)$ for every x such that
$\lim\limits_{t\to 0} g_x(t) = g_x(0) = 0$. Thus, $\sigma_n(x)$ converges to $f(x)$ at every point of
continuity of f—that is, in Lebesgue's terms (which Fejér did not use),
almost everywhere. This result, like the Fundamental Theorem I in
Riemann's theory, appears to depend on the fact that a Riemann-integrable function is continuous almost everywhere; but again Lebesgue
was able to extend Fejér's result to summable functions. Lebesgue was
able to relate Fejér's result to the Fundamental Theorem I for Lebesgue
integrals (Theorem 5.10) through his discovery that $\sigma_n(x)$ converges to
$f(x)$ for any x such that $F_x{}'(0) = 0$, where $F_x(u) = \displaystyle\int_0^u |\, g_x(t)\,|\, dt$. Thus
$\sigma_n(x)$ converges to $f(x)$ almost everywhere provided $F_x{}'(0) = 0$ almost
everywhere. Clearly it suffices to show that $\displaystyle\int_0^u |\, f(x+t) - f(x)\,|\, dt$

and $\displaystyle\int_0^u |\, f(x-t) - f(x)\,|\, dt$ have zero derivatives at $u = 0$ for almost
all x.

Making appropriate changes of variable in these integrals and replacing $f(x)$ by a, Lebesgue observed that the desired result follows from
the following:

> **THEOREM 6.2.** *If f is summable on $[0, 2\pi]$, then, for almost all
> values of x,* $\left(\displaystyle\int_0^x |\, f(t) - a\,|\, dt\right)' = |\, f(x) - a\,|$ *for every real num-
> ber a.*

Theorem 6.2 is a generalization of the Fundamental Theorem I, which
states that, for any fixed value of a, $\left(\displaystyle\int_0^x |\, f(t) - a\,|\, dt\right)' = |\, f(x) - a\,|$ except
on a set $E(a)$ of x of measure zero. But the fact that the union of the
$E(a)$ is likewise of measure zero does not follow immediately, since
there are an uncountable number of the $E(a)$. On the other hand, the
set E, which is the union of all those $E(a)$ such that a is rational, is of
zero measure, being a countable union of such sets. Hence Theorem

6.2 is true in the case that a is rational. To extend the theorem to irrational a, Lebesgue observed that the inequality

$$\left|\, |f(t) - a| \,-\, |f(t) - b| \,\right| \le |a - b|,$$

which is valid for any a and b, implies that

$$\left| \frac{1}{h} \int_x^{x+h} |f(t) - a| \, dt - \frac{1}{h} \int_x^{x+h} |f(t) - b| \, dt \right| \le |a - b|.$$

This inequality ensures that the validity of Theorem 6.2 for rational a implies its validity for irrational b, so the proof of Theorem 6.2 is complete.

Lebesgue's extension of Fejér's theorem also made possible a neat proof of Parseval's equality for bounded summable functions. Lebesgue's proof is contained in his second book, *Leçons sur les séries trigonométriques* [1906a], based on his lectures for the Cours Peccot at the College de France in the academic year 1904–05. Parseval's equality for Fourier series is

$$(5) \qquad \int_0^{2\pi} f^2 = \tfrac{1}{2} a_0 + \sum_{n=1}^\infty (a_n^2 + b_n^2),$$

where a_n and b_n are the Fourier coefficients of f. It was well known that the right-hand side converges to a limit less than or equal to the left-hand side. Harnack had attempted a proof that equality holds, but his proof was incorrect. A number of proofs were finally given in the late 1890's and early 1900's.[2] A. Hurwitz, who was chiefly responsible for making the validity of (5) known among mathematicians, at first actually doubted that (5) could in fact be true for functions as general as those within the scope of Riemann's theory of integration. His two proofs make crucial use of the fact that the set of points at which the oscillation of f exceeds any $k > 0$ has zero content.

Again, however, Lebesgue succeeded in demonstrating not only the fact that Riemann integrability was not a necessary condition but also the importance of his theorem on term-by-term integration. If f is bounded, then the sequence of functions $\sigma_n(x)^2$ is uniformly bounded and converges to $f(x)^2$ almost everywhere. Hence Theorem 5.3 is applicable, and

$$\int_0^{2\pi} f^2 = \lim_{n \to \infty} \int_0^{2\pi} \sigma_n^2 = \lim_{n \to \infty} \left[\tfrac{1}{2}a_0^2 + \sum_{k=1}^n (a_k^2 + b_k^2)\left(1 - \frac{k}{n}\right)^2 \right].$$

2. See [Hurwitz 1903]. Hurwitz overlooked a proof given by de la Vallée-Poussin [1893].

Parseval's equality then follows because it can easily be shown [cf. Lebesgue 1906a: 100–01] that the right-hand side of the equation is equal to $\frac{1}{2} a_0^2 + \sum\limits_{n=1}^{\infty} (a_n^2 + b_n^2)$.

Parseval's equality for any summable function, bounded or not, was obtained shortly thereafter by Pierre Fatou (1878–1929). As his doctoral thesis Fatou [1906] presented a paper on "Séries trigonométriques et séries de Taylor" with the principal aim "to show the advantage that can be obtained in these questions from the new notions of the measure of sets and the generalized definite integral" [1906: 335]. He was particularly encouraged in this undertaking by his "friend H. Lebesgue who has not ceased to interest himself with my researches and whose advice has been very useful to me" [1906: 338].

As the title indicates, Fatou's objective was to study the relationship between a summable function f and the associated harmonic function

$$u(r,t) = \tfrac{1}{2} a_0 + \sum_{n=1}^{\infty} (a_n \cos nt + b_n \sin nt) r^n,$$

where a_n and b_n are the Fourier coefficients of f. His interest in this matter may have been stimulated by Lebesgue's extension of du Bois-Reymond's theorem, for Lebesgue had suggested that the classical results on $u(r,t)$ can be extended to more general functions f. Fatou's main result is the following:

> THEOREM 6.3.   *If f is any function of period 2π which is summable and square-summable on $[-\pi,\pi]$, then $\lim\limits_{r \nearrow 1} u(r,t) = f(t)$ for almost all t.*[3]

This theorem generalizes a result of H. A. Schwarz [1872] for the case in which f is continuous. Actually Schwarz's proof makes it clear that when f is simply Riemann-integrable, $u(r,t)$ converges to $f(t)$ for every t that is a point of continuity of f, hence almost everywhere. In 1883 Harnack considered the case in which f is possibly unbounded but possesses an integral in accordance with his definition as given in 3.3. In fact, it was in order to deal with unbounded f that Harnack [1883: 320ff] introduced his extension of Riemann's integral. What he attempted to prove was that $u(r,t)$ converges to $f(t)$ "in general," but his proof is untenable when f is unbounded.

Fatou's proof of Theorem 6.3, which involves several lemmas that

---

3. Fatou actually showed his results are valid if $re^{it'}$ approaches e^{it} along any path that does not become tangent to the circle at e^{it}.

are important in their own right, and of the corollaries issuing from it demonstrated not only his talents as a mathematician but also the powerful methods of the new theory of integration. Throughout the proof of Theorem 6.3, Fatou made use of the well known fact that $u(r,t)$ has a Poisson integral representation,

$$u(r,t) = \frac{1}{2\pi} \int_{-\pi}^{\pi} P_r(u - t)f(u)du,$$

where

$$P_r(x) = \frac{1 - r^2}{1 + r^2 - 2r \cos x}.$$

The first step was to prove Theorem 6.3 for bounded, summable functions. This he accomplished with the help of the following proposition, derived by classical methods [1906: 345ff]:

THEOREM 6.4. *If* $U(r,t) = \dfrac{1}{2\pi} \displaystyle\int_{-\pi}^{\pi} P_r(u-t)F(u)du$, *then* $\lim\limits_{r \nearrow 1} \partial U/\partial t = F'(t)$ *for all* t *such that* $F'(t)$ *exists.*

Now suppose that f is bounded and summable. Then by Lebesgue's Theorem 5.10, $F'(t) = \left(\displaystyle\int_{-\pi}^{t} f \right)' = f(t)$ almost everywhere. Also without loss of generality it can be assumed that F is periodic. If the Poisson integral representation of $u(r,t)$ is integrated by parts and then Theorem 6.4 is applied,

(6) $\lim\limits_{r \nearrow 1} u(r,t) = \lim\limits_{r \nearrow 1} \partial U/\partial t = F'(t) = f(t)$ almost everywhere,

so Theorem 6.3 is proved for bounded functions.

The method by which Fatou chose to extend this result to unbounded functions, as he pointed out, is not the most direct but has the advantage of revealing some interesting side results. He began by assuming that f is summable and square-summable and nonnegative. The associated function $u(r,t)$ still exists and is harmonic for $r < 1$, so u is the real part of a function $g = u + iv$ analytic in the disc $|z| < 1$. Since

$$\frac{1}{2\pi} \int_{-\pi}^{\pi} |g(re^{it})|^2 dt = \tfrac{1}{2} a_0^2 + \sum_{n=1}^{\infty} (a_n^2 + b_n^2)r^{2n}$$

and since the series in the right-hand side of the equation remains convergent for $r = 1$, it follows that the left-hand side of the equation converges to $\frac{1}{2} a_0^2 + \sum\limits_{n=1}^{\infty} (a_n^2 + b_n^2)$ as $r \nearrow 1$. Therefore e, the set of t for which $\limsup\limits_{r \nearrow 1} |g(re^{it})| = \infty$, has zero measure.[4] Next he considered the function $g_1 = 1/(g+1) = 1/[(u+1)+iv]$, which is bounded for $|z| < 1$. To show that $\lim\limits_{r \nearrow 1} g_1(re^{it})$ exists almost everywhere, he used Theorem 6.4 to prove the following interesting result [1906: 366ff]:

> THEOREM 6.5 (Fatou's Theorem). *If f is bounded and analytic for $|z| < 1$, then $\lim\limits_{r \nearrow 1} f(re^{it})$ exists almost everywhere.*

Fatou's Theorem, together with the fact that the set e has measure zero, shows that $f_1(t) = \lim\limits_{r \nearrow 1} u(r,t)$ exists as a finite number almost everywhere; it remains to show that $f_1(t) = f(t)$ almost everywhere. Straightforward considerations show that $f_1(t) \geq f(t)$ almost everywhere, so it suffices to show that f_1 is summable and that $\displaystyle\int_{-\pi}^{\pi} (f_1 - f) \leq 0$. Fatou was thus led to prove another result of intrinsic interest [1906: 375]:

> THEOREM 6.6 (Fatou's Lemma). *If $\{f_n\}$ denotes a sequence of positive, bounded functions that are summable on $[a,b]$ and if $f(x) = \lim\limits_{n \to \infty} f_n(x)$ exists almost everywhere, then f is summable on $[a,b]$ and*
>
> $$\int_a^b f \leq \liminf_{n \to \infty} \int_a^b f_n,$$

4. Fatou simply remarked that this follows by a line of reasoning similar to that employed by Lebesgue to prove his Theorem 5.3 on term-by-term integration. Perhaps he had something similar to the following in mind: Let M denote a positive number, and E_r the set of all t such that $|g(re^{it})|^2 \geq M$. If $r_n \nearrow 1$ and $E_n = E_{r_n}$, then $e \subset \bigcup\limits_{k=1}^{\infty} \bigcap\limits_{n=k}^{\infty} E_n$, which is usually denoted by $\liminf\limits_{n \to \infty} E_n$. Now $m(e) \leq m(\liminf\limits_{n \to \infty} E_n) \leq \liminf\limits_{n \to \infty} m(E_n)$, and

$$m(E_n) \leq \int_0^{2\pi} |g(r_n e^{it})|^2 dt/M \leq \left[\tfrac{1}{2} a_0^2 + \sum_{n=1}^{\infty} (a_n^2 + b_n^2) \right]/M. \qquad \text{Letting}$$

$M \nearrow \infty$, one sees that $m(e) = 0$. The sets $\liminf\limits_{n \to \infty} E_n$ and $\limsup\limits_{n \to \infty} E_n$—the latter being by definition equal to $\bigcap\limits_{k=1}^{\infty} \bigcup\limits_{k=n}^{\infty} E_n$—had been introduced by Borel [1905: 18], who also established the inequalities $m(\liminf\limits_{n \to \infty} E_n) \leq \liminf\limits_{n \to \infty} m(E_n) \leq \limsup\limits_{n \to \infty} m(E_n) \leq m(\limsup\limits_{n \to \infty} E_n)$.

provided that the numbers $\int_a^b f_n$, $n=1, 2, 3, \cdots$, have a finite upper bound.

Theorem 6.6 was applied to the problem at hand as follows: the $u(r,t)$ are positive functions of t and converge almost everywhere to $f_1(t)$ as $r \nearrow 1$; also $\int_{-\pi}^{\pi} u(r,t)dt = a_0\pi$ is bounded in r. By Theorem 6.6, therefore, f_1 is summable, and $\int_{-\pi}^{\pi} f_1 \leq a_0\pi = \int_{-\pi}^{\pi} f$, so $\int_{-\pi}^{\pi} (f_1 - f) \leq 0$.

Theorem 6.6 is similar to the theorem of Beppo Levi referred to in 5.3. Fatou proved his theorem as follows: Let L denote a positive number, and consider the set E of all x such that $f(x) \leq L$. A sequence of functions g_n is constructed by setting $g_n(x) = f_n(x)$ if $f_n(x) \leq L$ and $g_n(x) = f(x)$ otherwise. Then for x in E, $0 \leq g_n(x) \leq L$, and $g_n(x)$ converges to $f(x)$. Lebesgue's Theorem 5.3 applies and shows that $\int_E f = \lim_{n \to \infty} \int_E g_n \leq \lim_{n \to \infty} \inf \int_a^b f$. The lemma then follows, since the right-hand side of the inequality is independent of L.[5]

Thus Theorem 6.3 is proved for any nonnegative function. Fatou was then able to establish the completely general case without any difficulty.

As an immediate corollary, Fatou obtained Parseval's equality without Lebesgue's restriction to bounded functions. If f is square-summable, Theorem 6.3 implies that $f(t)^2 = \lim_{r \nearrow 1} u(r,t)^2$ almost everywhere. Furthermore, $u(r,t)^2$, as functions of t with parameter r, satisfy the conditions of Fatou's Lemma (Theorem 6.6), so

$$\int_{-\pi}^{\pi} f^2 \leq \lim_{r \nearrow 1} \inf \int_{-\pi}^{\pi} u(r,t)^2 dt = \tfrac{1}{2} a_0^2 + \sum_{n=1}^{\infty} (a_n^2 + b_n^2).$$

That, together with the known reverse inequality, implies equality.

Fatou not only utilized techniques based on Lebesgue's versions of the Fundamental Theorem I and of term-by-term integration, but also was the first to see the usefulness of Lebesgue's Theorem 5.13. With this theorem, which states that a continuous function of bounded variation is differentiable almost everywhere, Fatou [1906: 379ff] proved:

5. It should be noted that this proof only requires that the f_n be summable (not bounded) and that $\lim_{n \to \infty} \inf \int_a^b f_n$ be finite.

THEOREM 6.7. *If* $\frac{1}{2} a_0 + \sum_{n=1}^{\infty} (a_n \cos nt + b_n \sin nt)$ *is any trig-onometric series such that* $\lim_{n \to \infty} na_n = \lim_{n \to \infty} nb_n = 0$, *then it converges almost everywhere.*

A theorem due to A. Pringsheim, Fatou observed, shows that with the above condition on the coefficients a_n and b_n, the series converges for a particular value of t if and only if $\lim_{r \nearrow 1} u(r,t)$ exists, where

$$(7) \qquad u(r,t) = \tfrac{1}{2} a_0 + \sum_{n=1}^{\infty} (a_n \cos nt + b_n \sin nt)r^n.$$

Fatou's Theorem 6.3 is not applicable here because $u(r,t)$ is not defined by a Fourier series corresponding to some square-summable function. Fatou's idea was to use Lebesgue's Theorem 5.13 to establish, in effect, the existence of such a function. (F. Riesz was able to use this idea to demonstrate the Riesz-Fischer Theorem.)

With this objective in mind, Fatou introduced the function

$$(8) \qquad U(r,t) = \sum_{n=1}^{\infty} \frac{1}{n} (a_n \sin nt - b_n \cos nt)r^n,$$

obtained by formally integrating (7) term by term. The condition on the a_n and b_n implies that the series on the right-hand side of (8) converges uniformly and absolutely. That makes U sufficiently well behaved that Schwarz's results apply, and

$$U(r,t) = \frac{1}{2\pi} \int_{-\pi}^{\pi} P_r(u - t)F(u)du,$$

where $F(u) = U(r,1)$. The function F is continuous and, as Fatou proved, of bounded variation. Hence by Lebesgue's Theorem 5.13, $f(t) = F'(t)$ exists almost everywhere. It then follows easily from Fatou's earlier results, Theorem 6.4 and (6), that $f(t) = \lim_{r \nearrow 1} u(r,t)$ almost everywhere. (For simplicity Fatou has assumed $a_0 = 0$.)

Other than Lebesgue, no one during the first decade of the twentieth century did more to demonstrate the value of the new theory of integration and to call it to the attention of a larger segment of the mathematical community than the Hungarian mathematician F. Riesz (1880–1956). The direction that his research took in this period was largely shaped by the combined influences of Hilbert's groundbreaking work on the theory of integral equations, the doctoral thesis of Maurice

Fréchet on abstract topological spaces,[6] and the works of Lebesgue and Fatou. Riesz first became acquainted with Lebesgue's definition of the integral when he read *Leçons sur les séries trigonométriques* in 1906 and was then induced to read Lebesgue's thesis and *Leçons sur l'intégration* as well. The same year he also read Fatou's paper. According to Riesz, "the idea and the courage to attempt to apply this notion of the integral to problems which interested me came to me through reading . . . the excellent memoir of Fatou . . ." [1949: 29].

The first application that Riesz made of Lebesgue's ideas involved the notion of a space of summable functions. In his thesis (1906), Fréchet had introduced the concept of what is now called an abstract metric space. As a concrete realization of this concept, Fréchet had considered the set of all functions continuous on a fixed, closed interval I with the distance between two functions f and g defined as the maximum of $|f(x)-g(x)|$ for x in I. Riesz [1906] observed that Fréchet's ideas, applied to this space of continuous functions, made possible an entirely different proof of a theorem due to Erhard Schmidt, one of Hilbert's students. Schmidt had proved that any orthogonal system of functions continuous on an interval must be countable.[7] The concept of an orthogonal system can, of course, be extended to any system of functions that can be integrated. Riesz pointed out that Schmidt's theorem can be generalized by using Fréchet's ideas; it is simply necessary to apply them to an appropriate metric space of more general functions. As an illustration, Riesz considered the class of functions bounded and summable on a Lebesgue-measurable set E with the distance between two functions defined to be $\left[\int_E (f-g)^2 \right]^{1/2}$. Then if the convention is made that all functions f such that $\int_E |f| = 0$ are regarded as identical with the function that is everywhere zero on E, a metric space is obtained. Here we have the origin of the notion of the L^p spaces later introduced and studied by Riesz.

The concept of a complete orthogonal system of functions had been introduced by D. Hilbert (1862–1943) in his work on integral equations. Hilbert showed that the analysis of the integral equation

(9) $$f(s) = \varphi(s) + \int_a^b K(s,t)\varphi(t)dt,$$

6. For a detailed discussion of the work of Hilbert and Fréchet, see [Bernkopf 1966].

7. Schmidt presented this theorem at a meeting of the Göttingen mathematical society in 1905. It was later published [Schmidt 1906].

in which φ is the unknown and f and K are continuous, can be reduced to that of an infinite system of linear equations

$$(10) \qquad \alpha_p = x_p + \sum_{q=1}^{\infty} \alpha_{pq} x_q \qquad p = 1, 2, 3, \cdots,$$

where the coefficients α_{pq} and α_p are defined in terms of a complete orthogonal system of functions φ_p that are continuous on $[a,b]$. If φ is a continuous solution of (9), then the "Fourier coefficients" of φ with respect to the system $\{\varphi_p\}$, viz., $a_p = \int_a^b \varphi \varphi_p$, represent a square-summable solution to (10). Conversely, if (10) has a unique square-summable solution $x_p = a_p$, for $p = 1, 2, 3, \cdots$, then a continuous solution φ to (9) can be constructed.

Hilbert's work suggested to Riesz [1907a; 1907b] the following problem: If a_p, for $p = 1, 2, 3, \cdots$, is a square-summable sequence, does there exist a function φ such that $a_p = \int_a^b \varphi \varphi_p$, for $p = 1, 2, 3, \cdots$? The motivation for this problem—as well as for the notion of a complete orthogonal system—also comes from the theory of trigonometric series: when the φ_p are sines and cosines, Fatou's version of Parseval's equality shows that a necessary condition for a sequence to be the Fourier coefficients of a square-summable function is that the sequence be square-summable. The question therefore naturally arises: Is the square-summability of a sequence also a sufficient condition? Before Lebesgue generalized the notion of an integrable function, it is doubtful that this question would have aroused much interest, since the possibility of a positive answer would have appeared unlikely. But Riesz was able to answer this question completely by proving the following:

THEOREM 6.8 (Riesz-Fischer Theorem). *If $\{\varphi_n\}$ is an orthonormal system of square-summable functions defined on $[a,b]$ and if $\{a_n\}$ is a sequence of real numbers, then there exists a square-summable function f defined on $[a,b]$ such that $a_n = \int_a^b f \varphi_n$, for $n = 1, 2, 3, \cdots$, if and only if $\sum_{n=1}^{\infty} a_n^2$ converges.*

The method of proof employed by Riesz involves first establishing the theorem for the special case in which $[a,b] = [0,2\pi]$ and the φ_n are the usual sines and cosines; the theorem is then extended to the general case by means of the theory of systems of linear equations in infinitely many unknowns. It is in establishing the theorem for the special case that Riesz made crucial use of the work of Lebesgue and Fatou. Let a_0, a_1, b_1, a_2, b_2, $\cdots$ denote a square-summable sequence of numbers. Riesz considered the function

$$F(t) = \sum_{n=1}^{\infty} \frac{1}{n} (a_n \sin nt - b_n \cos nt).$$

The right-hand side converges uniformly. (The convergence of $\sum_{n=1}^{\infty} (a_n{}^2 + b_n{}^2)$ implies by Schwarz's inequality[8] the convergence of the series with terms $|a_n|/n$ and $|b_n|/n$.) The function $F(t)$ is, of course, $U(1,t)$, where $U(r,t)$ is the function introduced by Fatou in his proof of Theorem 6.7. Riesz did not credit Fatou's paper for specifically suggesting the proof of Theorem 6.8, although that seems likely. Indeed, repeating Fatou's reasoning, Riesz observed that $F(t)$ is continuous and of bounded variation, so, by Lebesgue's Theorem 5.13, $F'(t)$ exists almost everywhere. Fatou's results—and Riesz referred to them specifically here—then show that $\lim_{r \nearrow 1} u(r,t) = f(t)$ almost everywhere, where $f(t) = \frac{1}{2} a_0 + F'(t)$ when $F'(t)$ exists and $f(t) = 0$ otherwise and $u(r,t)$ is defined as in (5). (As in Fatou's proof of Theorem 6.7, the relevant results are Theorem 6.4 and equation (6).)

To complete the proof, Riesz had to deduce, from the fact that $u(r,t)$ converges to $f(t)$, that f is square-summable and that the a_n and b_n are its Fourier coefficients. Lebesgue had had a similar problem in dealing with the series (1). As in that problem, equation (3) shows that it suffices to justify the use of term-by-term integration. In Riesz's case, $u(r,t)$ need not be uniformly bounded. But, as Fatou had pointed out, $\int_{-\pi}^{\pi} u(r,t)^2 dt \leq \frac{1}{2} a_0{}^2 + \sum_{n=1}^{\infty} (a_n{}^2 + b_n{}^2)$ for all r less than 1. Riesz thus applied the following lemma:

Let f_n be square-summable on a measurable set M. If $f_n(x)$ converges to $f(x)$ almost everywhere and if $\int_M f_n{}^2 \leq K^2$ for some fixed constant K, then f is square-summable, and $\int_M f = \lim_{n \to \infty} \int_M f_n$.

8. That is, $|\sum c_n d_n| < (\sum |c_n|^2)^{1/2} (\sum |d_n|^2)^{1/2}$. For integrals this becomes $\left| \int_E fg \right| < \left(\int_E |f|^2 \right)^{1/2} \left(\int_E |g|^2 \right)^{1/2}$

Riesz did not provide a proof of this lemma, but there is little doubt that he obtained it from Fatou's Lemma (Theorem 6.6). First of all, it follows immediately from Fatou's Lemma that f^2 is summable; Schwarz's inequality then implies that f is summable. (M is assumed to have finite measure.) The validity of term-by-term integration can also be deduced from Schwarz's inequality, which implies the inequality[9] $\int_M |f - f_n| \leq dm(M) + 2K[m(n,d)]^{1/2}$, where $m(n,d)$ denotes the measure of the set of x in M such that $|f_n(x) - f(x)| > d$.

Independently of Riesz and almost simultaneously, Ernst Fischer (1875–1959) also arrived at the same result [1907a]. Fischer had published an expository paper [1904] in which he gave two new proofs of Parseval's equality (for Riemann-integrable functions). That paper documents not only his interest in this equality but also his familiarity with Harnack's work on trigonometric series (3.1). It should be recalled that Harnack tried without success to deduce, from the fact that

$$\lim_{m,n \to \infty} \int_0^{2\pi} (S_n - S_m)^2 = 0,$$

the existence of a limit function $g(x) = \lim_{n \to \infty} S_n(x)$ "in general," where S_n denotes the nth partial sum of the Fourier series of an integrable function. It was probably Harnack's paper that led Fischer to introduce, albeit within the context of Lebesgue's theory, the notion of mean convergence. Mean convergence is defined in the following manner: Let Ω denote the class of square-summable functions defined on $[a,b]$, and suppose that f_n is a member of Ω for $n = 1, 2, 3, \cdots$. Then the sequence of functions f_n is said to converge in the mean if $\lim_{m,n \to \infty} \int_a^b (f_n - f_m)^2 = 0$. It is said to converge in the mean to f, a member of Ω, if

$$\lim_{n \to \infty} \int_a^b (f - f_n)^2 = 0.$$

Fischer's principal result is the following:

THEOREM 6.9. *If f_n converges in the mean, then there exists an f in Ω such that f_n converges in the mean to f.*

The Riesz-Fischer Theorem then follows as a corollary, since, given any square-summable sequence a_n, the functions $f_n = \sum_{i=1}^n a_i \phi_i$ converge in the mean.

9. This inequality appears in a later paper by Riesz [1910: 463–64].

To prove the main theorem, Fischer first observed that Schwarz's inequality shows that $F(x) = \lim\limits_{n \to \infty} \int_a^x f_n$ and $\theta(x) = \lim\limits_{n \to \infty} \int_a^x f_n{}^2$ exist and are continuous. As θ is monotonic, it is of bounded variation. And, by another application of Schwarz's inequality,

$$| F(x+h) - F(x) |^2 \le h[\theta(x + h) - \theta(x)],$$

where h is positive. Thus if h_n is any sequence of positive numbers,

$$[\sum | F(x_n + h_n) - F(x_n) |]^2 \le (\sum h_n)^2[\theta(b)]^2,$$

where the x_n are in $[a,b]$; therefore, F is absolutely continuous (Fischer does not use this term and may not have read Vitali's paper). Fischer considered the functions f and ζ which, by definition, take the values $D^+F(x)$ and $D^+\theta(x)$, respectively, when these are finite and the value 0 otherwise. Both are summable (as can be seen from Lebesgue's proof of Theorem 5.11) and $\int_a^x f = F(x) = \lim\limits_{n \to \infty} \int_a^b f_n$ (by Theorem 5.15). The proof is completed by showing that $f^2(x) = \zeta(x)$ almost everywhere, so f belongs to Ω and f_n converges in the mean to f.

To emphasize that "use of the notions of M. Lebesgue . . . is necessary for our subject," Fischer pointed out in a subsequent note [1907b: 1151] that if Ω is replaced by the set of continuous functions on $[a,b]$ or by some other proper subset of Ω, his main theorem is no longer valid. In modern terms, the space Ω is complete with respect to the metric $d(f_1,f_2) = \left[\int_a^b (f_1 - f_2)^2 \right]^{1/2}$, while the other spaces are not. It is historically interesting that this property of the metric space Ω (that is, $L^2[a,b]$) was originally derived from Lebesgue's theorems on the differentiation of functions of bounded variation.

Underlying the discussions of term-by-term integration by Kronecker, Harnack, Arzelà, and Lebesgue had been the notion of convergence in measure, a notion first explicitly defined by Riesz [1909a]. (A sequence of measurable functions f_n is said to converge in measure to the function f if $\lim\limits_{n \to \infty} m(n,\epsilon) = 0$ for every $\epsilon > 0$, where $m(n,\epsilon)$ denotes the measure of the set of x such that $|f(x) - f_n(x)| > \epsilon$.) If f_n converges to f in measure, it need not converge to f pointwise; but Riesz showed that there is always a subsequence f_{n_k} of f_n such that $f(x) = \lim\limits_{k \to \infty} f_{n_k}(x)$ almost everywhere. A similar result for mean convergence was obtained independently by Hermann Weyl [1909: 243].

Riesz also pointed out that because convergence in the mean implies convergence in measure, the trigonometric series of a square-summable function contains a subseries that converges to the function almost everywhere. In a sense, that result together with Fischer's Theorem 6.8 showed the manner in which Harnack's intuitions could be given a valid interpretation.

Riesz's early contributions to the theory of integration culminated in a brilliant paper [1910] in which he introduced and studied the spaces $L^p[a,b]$, for $1 < p < \infty$, consisting of all measurable functions defined on $[a,b]$ such that $|f|^p$ is summable. The case in which $p = 2$, of course, had already been discussed by Riesz and Fischer. From a purely mathematical standpoint, the importance of Riesz's paper lies in its contribution to the development of the theory of integral equations and function spaces, and for this reason it will not be discussed here.[10] Historically, however, the paper is extremely important because it secured for the new theory of integration a permanent place in the development of the theory of function spaces. The crucial property of the L^p spaces not shared by the corresponding spaces of Riemann-integrable functions is that of completeness in the metric $d(f_1, f_2) =$

$$\left(\int_a^b |f_1 - f_2|^p \right)^{1/p}$$

10.   See [Bernkopf 1966] for a detailed analysis of its contents.

The Lebesgue-Stieltjes Integral

ALTHOUGH LEBESGUE'S work on measure and integration was concerned specifically with his own definition of measure and the resulting theory of integration, his ideas have been embodied in contemporary abstract theories of measure and integration. The first step towards these modern theories was taken by Johann Radon (1887–1956). Radon [1913] initiated the study and application of integrals based upon measures more abstract than Lebesgue's, frequently referred to as Lebesgue-Stieltjes integrals because the definitions of Lebesgue and Stieltjes are included as special cases. Not only did Radon's paper introduce a more abstract approach to the notion of a measure, it also suggested to others the possibility of further abstraction in the direction of removing the customary restriction to n-dimensional Euclidean space (see, e.g., [Fréchet 1915]).

The principal developments leading up to Radon's paper form a fitting epilogue to our study of Lebesgue's work and its historical context because they form the link between the generalization of the integral proposed by Lebesgue in 1901 and contemporary abstract treatments of measure and integration. Moreover, Lebesgue himself figured prominently in these developments.

The motivation for T.-J. Stieltjes' (1856–94) generalization of Cauchy's definition of the integral was not at all like Riemann's or Lebesgue's. Stieltjes [1894] set himself the task of making a general study of continued fractions with positive coefficients a_n:

179

(1)
$$a_1z + \cfrac{1}{a_2 + \cfrac{1}{a_3z + \cfrac{1}{a_4 + \cfrac{1}{a_5z + \cfrac{1}{\ddots}}}}}$$

He succeeded in showing that when the series Σa_n diverges, the continued fraction (1) converges to a limit $F(z)$ that is analytic in the plane less the negative real axis and the origin. And when Σa_n converges, he was able to prove, the even and odd partial sums of (1)—that is, the finite continued fractions ending with the coefficients a_{2n} and a_{2n-1}, respectively—converge to distinct limits $F_1(z)$ and $F_2(z)$ that can be represented in the forms

(2)
$$F_1(z) = \frac{\mu_1}{z + \lambda_1} + \frac{\mu_2}{z + \lambda_2} + \cdots + \frac{\mu_i}{z + \lambda_i} + \cdots;$$

$$F_2(z) = \frac{\nu_0}{z} + \frac{\nu_1}{z + \theta_1} + \cdots + \frac{\nu_i}{z + \theta_i} + \cdots;$$

with μ_i, λ_i, ν_i, θ_i, nonnegative.

It was well known that the continued fraction (1) could be formally developed in a series of the form

(3)
$$\frac{c_0}{z} - \frac{c_1}{z^2} + \cdots + (-1)^k \frac{c_k}{z^k} + \cdots.$$

with positive coefficients, and conversely. Stieltjes observed that when the series $\sum a_n$ converges, the coefficients c_k are related to F_1 and F_2 by the equalities

(4)
$$c_k = \sum_{i=1}^{\infty} \mu_i \lambda_i^k, \qquad k = 1, 2, 3, \cdots;$$

$$c_k = \sum_{i=0}^{\infty} \nu_i \theta_i^k, \qquad k = 0, 1, 2, \cdots.$$

These relationships suggested the following physical interpretation [Stieltjes 1894: J.48]: Consider a distribution of positive mass on the positive x-axis with mass m_i concentrated at a distance d_i from the origin. The sum $\sum m_i d_i{}^k$ is then defined to be the moment of order k of the system with respect to the origin. The formulae in (4) can then be interpreted to mean that c_k is the moment of order k of the system $m_i = \mu_i,\ d_i = \lambda_i$ or of the system $m_i = \nu_i,\ d_i = \theta_i$. Thus the "problem of moments"—which Stieltjes defined as the problem of finding a distribution of positive mass $m_i,\ d_i$ such that its moment of order k has a prescribed value c_k for each k—is solvable when the given numbers c_k have the property that the a_k, the coefficients of the continued fraction (1) associated with the series (3) form a convergent series. In this case the problem of moments has two and consequently infinitely many solutions.

In order to deal with the problem of moments when the a_k's associated with the c_k's form a divergent series, Stieltjes introduced the notion of a continuous distribution of mass [1894: J.68]. Such a distribution is completely determined when $g(x)$, the total mass between 0 and x, is known for every positive value of x. The function g is then a positive, increasing function of x. The moment of order k is defined analogously as the limit of the sums

$$\sum_{i=1}^{n} t_i{}^k [g(x_i) - g(x_{i-1})]$$

as $\|P\| \to 0$, where $P,\ a = x_0 < \cdots < x_n = b$, is a partition of $[a,b]$ and t_i is in $[x_{i-1}, x_i]$. Stieltjes showed that when f is any continuous function and g is not necessarily positive, the corresponding limit of

$$\sum_{i=1}^{n} f(t_i) [g(x_i) - g(x_{i-1})]$$

still exists, and he denoted it by $\displaystyle\int_a^b f(u) dg(u)$. Stieltjes realized that his definition could be extended to a larger class of functions, but he felt that "it is of no interest to accord complete generality to the function $f(u)$" [1894: J.71].

The problem of moments thus became: Given positive numbers c_k, find a positive, nondecreasing function g such that $c_k = \displaystyle\int_a^b u^k dg(u)$.

The problem posed, Stieltjes was then able to completely resolve it by showing that when Σa_k, the series corresponding to the numbers c_k, diverges, there exists a unique solution g that is related to $F(z)$, the limit of (1), by

$$F(z) = \int_0^\infty \frac{dg(u)}{z + u}.$$

Similarly, when Σa_n converges, two different solutions g_1 and g_2 can be determined and satisfy

$$F_1(z) = \int_0^\infty \frac{dg_1(u)}{z+u} \text{ and } F_2(z) = \int_0^\infty \frac{dg_2(u)}{z+u},$$

the integrals in this case reducing to the expressions in (2).

Stieltjes' definition of the integral continued to remain outside the tradition stemming from Riemann's paper until 1910. By that time a number of mathematicians, mostly Italian and French, had become interested in the study of what are today usually called "functionals," that is, correspondences $f \rightarrow U(f)$ between functions f and real or complex numbers $U(f)$. Jacques Hadamard (1865–1963) showed [1903] that if U is a linear functional defined for all continuous functions on $[a,b]$ and continuous in the sense that $U(f_n) \rightarrow U(f)$ whenever f_n converges uniformly to f, then U can always be represented in the form

$$U(f) = \lim_{u \to \infty} \int_a^b f(t)G(t,u)dt,$$

where $G(t,u)$ is the value of U for the function $x \rightarrow (u/\pi^{1/2})e^{-u^2(t-x)^2}$. This representation followed easily from applying U to the well known formula

$$f(x) = \lim_{u \to \infty} u/\pi^{1/2} \int_a^b f(t)e^{-u^2(t-x)^2}dt.$$

A more profound and satisfactory representation of U was obtained later by F. Riesz [1909b]. Riesz showed that if Stieltjes' original definition of $\int_a^b f(u)dg(u)$ is extended to functions g of bounded variation on $[a,b]$—an extension which is immediate by virtue of Jordan's representation of these functions as the difference of increasing functions—then a function of bounded variation g can be determined so that for all f,

$$U(f) = \int_a^b f(u)dg(u).$$

It was Riesz's paper that turned Lebesgue's interest to the Stieltjes integral; in a brief note Lebesgue [1910a] proposed to indicate "the close connections" between it and his own definition. From a heuristic standpoint the natural relationship would be $\int_a^b f(x)dg(x) =$

$\int_a^b f(x)g'(x)dx$. And that is essentially what Lebesgue confirmed, although since g need not be continuous or absolutely continuous, he had to resort to a change of variable similar to that used in his study of the integral formula for arc length (5.1). Thus if $v = v(x)$ denotes the total variation of g on $[a,x]$, the idea is to construct an inverse function defined on $[0,v(b)]$ with range $[a,b]$. When v is continuous and strictly increasing, the inverse function $v^{-1}(v) = x(v)$ exists in the ordinary sense; when v has intervals of invariability and jump discontinuities, an inverse, possibly discontinuous, can still be defined by making the obvious modifications. If g is continuous, then the function $h(v) = g(x(v))$ is continuous and has total variation equal to v on $[0,v]$. Again when g is not continuous, the above definition can be suitably modified so that the resulting h continues to possess those properties. Thus h is absolutely continuous and is an indefinite integral

$$h(v) = \int_0^v k(t)dt; \text{and it is possible to verify the equality}$$

$$\int_a^b f(x)dg(x) = \int_0^{v(b)} f(x(v))dh(v) = \int_0^{v(b)} f(x(v))k(v)dv.$$

The Stieltjes integral of a continuous function f can consequently be identified with a Lebesgue integral that is unique in the sense that g determines $x(v)$ and $k(v)$ almost everywhere on $[0,v(b)]$. The third term in the above equality, since it is a Lebesgue integral, retains a meaning when f is not continuous; and Lebesgue proposed to generalize Stieltjes' definition by defining $\int_a^b f(x)dg(x)$ to be $\int_0^{v(b)} f(x(v))k(v)dv$ whenever $f(x(v))$ is bounded and summable. It seemed to Lebesgue that it would be difficult to obtain a generalization of the Stieltjes integral that did not involve such a change of variables. Later the same year, however, Lebesgue published a paper [1910b] in which he extended his results on the theory of integration to n-dimensional space by means of a point of view that was to suggest a more natural method of bringing out "the close connections" between the Lebesgue and Stieltjes integrals.

Lebesgue's paper [1910b] was preceded by, and influenced by, a paper by Vitali [1907–08]. Although Lebesgue's definition of the integral can be extended to functions defined on points of n-dimensional space by simply replacing 1-dimensional by n-dimensional points, that cannot be done for the theorems involving derivatives and indefinite integrals and for such methods used to prove them as the method of chains of intervals. Vitali [1907–08] was the first to concern himself with these matters. He was particularly interested in the problem of extending the meaning and the proof of his own theorem that every absolutely continuous function is an indefinite integral. The major portion of Vitali's paper is concerned with the problem of re-proving this theorem by a method that would generalize to n-dimensional space. In the last section Vitali considered the meaning to be given to the terms "absolutely continuous" and "indefinite integral" for functions of several variables, leaving the reader to justify the theorem identifying absolutely continuous functions and indefinite integrals by imitating the proof for one dimension. For two dimensions, he proceeded as follows: Suppose $F(x,y)$ is defined on a rectangle R_0 determined by $0 \leq x \leq A$, $0 \leq y \leq B$. We shall use the notation $[0,A;0,B]$ to refer to R_0 and similar notation for other rectangles with sides parallel to the axes. Let $R = [a,b;c,d]$ be contained in R_0. Vitali defined the increment of F in R to be

$$F(a,c) + F(b,d) - F(a,d) - F(b,c).$$

Then if $R_n = [a_n,b_n;c_n,d_n]$, for $n = 1$, 2, 3, $\cdots$, is a sequence of non-overlapping rectangles, the increment of F in $\bigcup_{n=1}^{\infty} R_n$ is taken to be the sum of the increments of F in each R_n. And F is defined to be absolutely continuous if the increment of F in $\bigcup_{n=1}^{\infty} R_n$ approaches zero with the area of $\bigcup_{n=1}^{\infty} R_n$. Finally, F is called an integral function if there exists a summable function $f(x,y)$ defined on R_0 such that

$$F(x,y) + F(0,0) - F(x,0) - F(0,y) = \int_{R(x,y)} f(x,y),$$

where $R(x,y) = [0,x;0,y]$ is contained in R_0. In other words, the increment of F in $R(x,y)$ is the integral of f over $R(x,y)$.

Although Vitali's definitions take the notion of point function as fundamental, implicit in them is the notion of a set function, specifically, a function defined on rectangles. And it is this notion that Lebesgue [1910b] decided to take as fundamental in his own investigations. It also appears that he was influenced in this direction by the work of Volterra, who in 1887 initiated the interest in functionals referred to earlier. The starting point for Volterra's work was his recognition that Dirichlet's conception of a function as a correspondence can be extended immediately to the case in which the variable is not a number or a point in space but is itself a function. In a paper summarizing some of his work, Volterra [1889] expressed his point of view in the following manner:

> The representation of functions of three independent variables by *functions of the points of a three-dimensional space* is a very widespread practice among analysts. But points are not the only geometrical elements of space. There are also lines and surfaces and, in the same manner, the values of a variable can be made to correspond to each point or to each line or to each surface. In this manner *functions of the points* of space are obtained and also what can be called *functions of the lines* and *functions of the surfaces*. Until now analysis has only been applied to functions of points, but it is also very interesting to study functions of lines and functions of surfaces. [1889: 233.]

In full agreement with Volterra's viewpoint, with which he was familiar, Lebesgue observed that the indefinite integral can be considered quite naturally as a function of measurable sets, viz.

$$F(E) = \int_E f(P)dP,$$ where P is a point in n-dimensional space [1910b: 380ff]. Peano had made this observation much earlier (4.1), but it appears that Lebesgue was not acquainted with Peano's ideas on the matter. Lebesgue singled out two important properties of this class of set functions:

(a)   If $m(E_n) \to 0$, then $F(E_n) \to 0$;

(b)   If $E_1, E_2, E_3, \cdots$ *are pairwise disjoint, measurable sets, then*

$$F\left(\bigcup_{n=1}^{\infty} E_n\right) = \sum_{n=1}^{\infty} F(E_n).$$

An arbitrary set function defined on measurable sets is then defined to be absolutely continuous if (a) holds and to be additive if (b) holds.

In order to establish the analogue of his and Vitali's theorem on absolutely continuous functions, it was necessary to develop a theory of differentiation for set functions. A detailed discussion of this theory would take us too far afield from our principal objective, which is to trace the events leading up to the creation of the theory of the Lebesgue-Stieltjes integral by Radon. Let it suffice to say that Lebesgue found the type of definition of differentiability given by Peano to be too restrictive, since, for example, the Fundamental

Theorem I would not be valid for $F(E) = \int_E f(P)dP$. But Lebesgue found that if he restricted the set E that occurs E in the quotient $\dfrac{F(E)}{m(E)}$ to certain "regular families" of sets containing P—for example, the family of spheres with center P—then he obtained a definition of the derivative (or of the Dini derivatives) of F at P for which the propositions for functions of a single variable remain true. Thus Lebesgue was able to demonstrate the following theorem:

> **THEOREM 1.** *If $F(E)$ is absolutely continuous and additive, then F possesses a finite derivative almost everywhere. Furthermore,*
>
> $$F(E) = \int_E f(P)dP,$$ *where $f(P)$ is equal to the derivative of F at P when this exists and $f(P)$ is equal to arbitrarily chosen values otherwise.*

In his discussion of the indefinite integral $F(E) = \int_E f(P)dP$, Lebesgue had also pointed out that F is of bounded variation on E, by which he meant the following: Let E be partitioned into pairwise disjoint, measurable sets $E_1, E_2, E_3, \cdots$, and let $P(E)$ and $N(E)$ denote the least upper bounds of the numbers $\Sigma'F(E_n)$ and $-\Sigma''F(E_n)$, respectively, where Σ' and Σ'' denote summation over all the terms $F(E_n)$ which are positive and negative, respectively. $P(E)$ and $N(E)$ are termed the positive and negative variations of F on E, and $T(E) = P(E) + N(E)$ is defined to be the total variation. F is of bounded variation on E when its total variation on E is finite. These definitions are clearly applicable to any set function F, although Lebesgue gives them in the context of indefinite integrals. He also observed in passing (and without proof) that any additive set function $F(E)$ is automatically of bounded variation, because the series $\sum_{n=1}^{\infty} F(E_n)$ corresponding to $E = \bigcup_{n=1}^{\infty} E_n$ must always converge absolutely.

This observation on Lebesgue's part is extremely important because it indicated to Radon [1913] that the concept of an additive set function could be used to give a definition of the integral comprehending both Lebesgue's and Stieltjes'. Radon's motivation for such an extension of the integral came from Hilbert's theory of integral equations. The Stieltjes integral had been employed by Hilbert to obtain the canonical representation of a bounded quadratic form. On the other hand, Riesz's 1910 paper on L^p spaces had made fundamental use of the Lebesgue integral. With the introduction of a definition of the integral that included these definitions as special cases, the groundwork would be laid for a general theory.

Radon defined an interval in n-dimensional space to be a set of points $P = (x_1, x_2, \cdots, x_n)$ satisfying inequalities of the form $a_i \leq x_i < b_i$, for $i = 1, 2, \cdots, n$. All sets will be understood to be subsets of the interval J defined by the inequalities $-M \leq x_i < M$, for $i = 1, 2, \cdots, n$. Radon considered a class of sets T satisfying the following conditions:

(a)   *All intervals belong to T;*

(b)   *If E_1 and E_2 belong to T, so do $E_1 \cap E_2$ and $E_1 - E_2$;*

(c)   *If E_1, E_2, E_3, $\cdots$ belong to T and are pairwise disjoint, then*

$$\bigcup_{n=1}^{\infty} E_n \text{ is also in } T.$$

These conditions imply that all Borel sets are included in T. A set function f defined on T is defined to be additive if $f\left(\bigcup_{n=1}^{\infty} E_n \right) = \sum_{n=1}^{\infty} f(E_n)$ whenever the E_n are pairwise disjoint and belong to T. It may happen that f satisfies the additional important condition:

(d)   *Given any E in T and any $\epsilon > 0$, there exists a closed set E' contained in E such that $|f(E) - f(E')| < \epsilon$.*

In accordance with Lebesgue's observation, Radon showed that if f is additive, then f is of bounded variation and can be represented as the difference of monotone (i.e., nonnegative) additive set functions; that is, for all E in T,

(5) $$f(E) = \varphi(E) - \psi(E).$$

Radon's first objective was to "complete" the domain of definition of f in a manner analogous to Lebesgue's completion of Borel's definition of measure. The representation (5) shows that it suffices to consider such a completion for monotone f. Assuming at first that f is monotone, Radon introduced the analogues of inner and outer measures. Given an arbitrary set E, then $\bar{f}(E)$ is defined to be the greatest lower bound of numbers of the form $\sum_i f(J_i)$, where the J_i denote intervals whose union contains E; and $\underline{f}(E)$ is defined to be $f(J) - \bar{f}(J - E)$. E is said to be measurable with respect to f if $\underline{f}(E) = \bar{f}(E)$. Radon showed that the class T_1 of all f-measurable sets satisfies conditions (a), (b), and (c). Hence f may be defined on T_1 by setting $f(E)$ equal to the common value of $\underline{f}(E)$ and $\bar{f}(E)$. Furthermore, if the original f and T satisfy (d), then T is contained in T_1 and f defined on T_1 is a bona fide extension of f to a larger class of sets. T_1 is called the natural domain of definition of f.

When f is not monotone, the extension procedure is applied to φ and ψ of (5), and the natural domain of f is defined as the intersection of the natural domains of φ and ψ. If f and T satisfy (d), so do φ and ψ with respect to T, so again the natural domain of f contains T. Finally, Radon showed that the procedure used to extend T to the larger natural domain T_1 yields nothing new if it is now applied to T_1.

Once these notions have been established, the generalization of the integral is automatic. A function F is measurable with respect to f if, for every real number a, T_1 contains the set of points P such that

$F(P) > a$; a measurable function F is summable if the series $\displaystyle\sum_{k=-\infty}^{\infty} a_k f(E_k)$

converges absolutely, where $\cdots < a_{-2} < a_{-1} < a_0 < a_1 < a_2 < \cdots$ is a partition of $(-\infty, \infty)$ with finite norm and E_k denotes the set of points P satisfying $a_k \leq F(P) < a_{k+1}$. Finally, when F is summable with re-

spect to f, $\displaystyle\int_J F(P)df$ is defined to be the limit of the above series as

the norm of the partition approaches zero. Because these are precisely Lebesgue's definitions with m replaced by f, Radon's definition contains Lebesgue's as a special case. It also contains Stieltjes' definition: Let F be continuous and G of bounded variation on $[a,b]$. Without loss of generality it can be assumed that G is nondecreasing and that $G(x) = G(x-)$ for all x.[1] In Radon's definition of T_1 where f is monotone, f need only be initially defined on intervals and satisfy the condition that $f(J_0) = \displaystyle\sum_i f(J_i)$ whenever the interval J_0 is expressed as a

union of disjoint intervals J_i. From the point function G, such an f can be defined by setting $f([a,b)) = G(b) - G(a)$; the left-hand continuity of G implies that f has the additivity property on intervals.[2] The set functions $\underline{f}$ and $\bar{f}$ can then be defined and f can be extended to the class T_1 of f-measurable sets. If F is continuous on the inter-

---

1. If $G^*(x) = G(x-)$, then $\displaystyle\int_a^b F(u)dG(u) = \int_a^b F(u)dG^*(u)$.

2. Cf. Radon's remarks [1913: 1305ff].

val $[-M,M]$, then, it is not difficult to see, $\displaystyle\int_{-M}^{M} F df$ is precisely $\displaystyle\int_{-M}^{M} F(x)dG(x)$.

From the standpoint of the theory of integration, Radon's most significant result is his generalization of Lebesgue's Theorem 1 on absolutely continuous set functions, which in its abstract form is commonly referred to as the Radon-Nikodym Theorem. To begin with, the notion of absolute continuity is dissociated from that of Lebesgue measure. Given two additive set functions b and f with natural domains of definition T_b and T_f, then b is defined to be a basis for f if b is nonnegative and has the property that if $b(E)=0$ for any set E in $T_b \cap T_f$, then $f(E)=0$. (This is essentially the notion of coexistent set functions introduced by Cauchy and later by Peano; see 4.1.) Using (d), Radon proved that T_b is contained in T_f and that for every $\epsilon > 0$ there exists a $\delta > 0$ such that $|f(E)| < \epsilon$ whenever $b(E) < \delta$. Thus, when b signifies Lebesgue measure, f has b as a basis when f is absolutely continuous.

Radon was able to generalize Theorem 1 to:

THEOREM 2. *If g is an additive set function with basis f, then there exists an f-summable function Ψ such that* $g(E) = \displaystyle\int_E \Psi(P)df$ *for every E in T_f.*

The method Radon employed to establish this result has a rather interesting history and illustrates the manner in which an idea can evolve to acquire a significance and utility that bear little resemblance to the original intentions. The method consists in constructing a certain "measure-preserving" correspondence between T_f and one-dimensional, Lebesgue-measurable subsets of the interval $K = [0, f(J))$; thus it is possible to deduce Theorem 2 from Lebesgue's Theorem 1.

The basic idea behind the construction of this correspondence can be traced back to Peano's discovery of space-filling curves (mentioned in 4.1). Peano had defined his example arithmetically. Shortly thereafter Hilbert [1891] observed that the type of curve considered by Peano could be given a more intuitive construction as follows:[3] To define a continuous mapping of [0,1] onto the unit square [0,1;0,1], divide the interval into four equal intervals K_1, K_2, K_3, K_4 and the

---

3. In our presentation a different notation is used in order to make the relationship to Radon's proof clear.

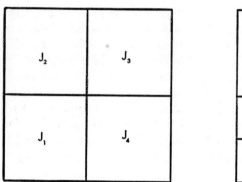 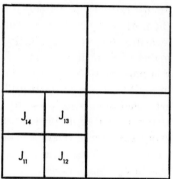

Fig. 7

square into four equal squares J_1, J_2, J_3, J_4; associate each K_i with a J_i, as indicated in the left-hand side of Figure 7. Then K_1 is divided into four equal intervals K_{11}, K_{12}, K_{13}, K_{14}, and J_1 into four corresponding squares J_{11}, J_{12}, J_{13}, J_{14}, as indicated in the right-hand side of Figure 7. The same process of subdivision is repeated in K_2 and J_2, in K_3 and J_3, in K_4 and J_4, and so on. At the nth stage in this process the unit interval will be divided into 4^n equal intervals $K_{m_1\cdots m_n}$ and the unit square into 4^n corresponding squares $J_{m_1\cdots m_1}$, where each m_i is equal to 1, 2, 3, or 4. Thus, if Q is a point of [0,1], Q will belong to at least one of the corresponding intervals $K_{m_1\cdots m_n}$ for each n. In fact, for each n, the interval $K^Q_{m_1\cdots m_n}$ that contains Q can be chosen in such a manner that $K^Q_{m_1\cdots m_n m_{n+1}}$ is contained in $K^Q_{m_1\cdots m_n}$ and that Q is the only point common to all these intervals. Furthermore, the correspondence between squares and intervals was determined so that the squares $J^Q_{m_1\cdots m_n}$ corresponding to the intervals $K^Q_{m_1\cdots m_n}$ also form a nested sequence. Hence, $\bigcap_{n=1}^{\infty} J^Q_{m_1\cdots m_n}$ contains a single point P. The correspondence $Q{\rightarrow}P$ is a continuous mapping of the unit interval onto the unit square. Figure 8, which is taken from Hilbert's paper, indicates how the successive subdivisions of the square can be used to obtain successive approximations to the space-filling curve.

As early as 1899 Lebesgue had stressed the usefulness of the space-filling curves of Peano and Hilbert for extending results known to hold for the real line to higher dimensional space [1899a: 812]. And in his *Leçons sur l'intégration* he demonstrated the Heine-Borel Theorem for two dimensions by this method to illustrate "how the curve of M. Peano and the other analogous curves can be used . . . " [1904: 117]. Lebesgue also explained [1910b] that he had originally intended to extend his results on integration theory to functions of several variables by means of the correspondence between n-dimensional space and one-dimensional space determined by these curves. It was undoubtedly that correspondence that F. Riesz [1910] also had in mind when he

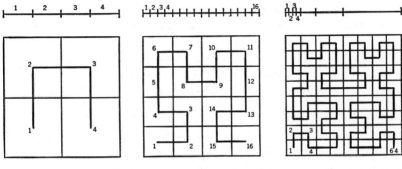

Fig. 8

observed that most of his results for $L^p(M)$ with $M = [a,b]$ could be established for an arbitrary measurable subset M of n-dimensional space by applying the following correspondence principle:

> *If a measurable set M of arbitrary dimension and of measure m is given, this set can be mapped essentially biuniquely onto a line of length m in such a manner that to every measurable subset of the one set corresponds a measurable subset of the other and with the same measure.* The expression "essentially biuniquely" is to be explained in the following manner: For each of the two sets, those points with which no point, or more than one point, is associated in the other set form a null set. [1910: 497.]

By employing a method that is essentially Hilbert's, Radon explicitly defined Riesz's correspondence and extended it to cover the case in which n-dimensional Lebesgue measure is replaced by the set function f and measurable sets are replaced by T_f.[4] In its extended form the principle becomes:

THEOREM 3. *There exists a subset E_0 of J of f-measure 0 and a subset $\overline{E}_0$ of $K = [0, f(J))$ of Lebesgue measure 0 and mappings α and β defined on the points of $J - E_0$ and $K - \overline{E}_0$, respectively, with the following properties:*

(i) *Except for at most a countable number of points $P = P_i$, $\alpha(P)$ is a point Q of K. The mapping α associates with each exceptional point P_i an interval $A_i = \alpha(P_i)$ in such a manner that if $i \neq i'$, then $A_i \cap A_{i'} = \varnothing$.*

(ii) *$\beta(Q)$ is a point of $J - E_0$ for each Q in $K - \overline{E}_0$ and $\alpha(\beta(Q)) = Q$ if Q does not belong to one of the intervals A_i.*

(iii) *Let E be any f-measurable subset of J and let E' denote the set of points Q in K such that either $\alpha(P) = Q$ for some P in E or Q belongs to A_i and P_i is in E. Then E' is measurable and $m(E') = f(E)$, where m denotes Lebesgue measure.*

(iv) *If E' is any measurable subset of K and E denotes the set of P in J such that $P = \beta(Q)$ for some Q in $E' - \bigcup_i A_i$, then E is f-measurable and $f(E) = m(E')$.*

In order to indicate the close connection between this proposition and Hilbert's construction of the space-filling curve, Radon's proof of (i) and (ii) will be sketched for the case in which J is a two-dimensional interval, that is, a square with certain edges not included. Suppose J is successively divided into squares as in Hilbert's construction—the only difference being that now the squares of a particular subdivision are pairwise disjoint, since all squares are taken to be intervals in Radon's sense. The corresponding subdivision of the interval $K = [0, f(J))$ is modified by associating with J_1, J_2, J_3, J_4 intervals K_1, K_2, K_3, K_4 that are of lengths $f(J_1)$, $f(J_2)$, $f(J_3)$, $f(J_4)$, respectively, rather than of lengths $(1/4)f(J)$. Similarly, at the next stage of the subdivision, the intervals K_{11}, K_{12}, K_{13}, K_{14} corresponding to J_{11}, J_{12}, J_{13}, J_{14} are taken with lengths $f(J_{11})$, $f(J_{12})$, $f(J_{13})$, $f(J_{14})$, respectively, and so on.

4.   Radon only referred to the above principle of Riesz's, but there is little doubt that Hilbert's well known construction inspired his proof. For example, Lebesgue [1910b] discussed it in a paper Radon had studied. In fact, Lebesgue [1910b: 401–04] discussed the idea of using Hilbert's space-filling curve to prove his Theorem 1 by reducing it to the one-dimensional case (Theorem 5.15). Radon's proof of Theorem 2 is based on this idea.

As in Hilbert's construction, every P in J determines a nested sequence of squares $J^P_{m_1 \cdots m_n}$, for $n = 1, 2, 3, \cdots$ such that P is the unique element in their intersection; since the squares are intervals, there is only one such sequence. Let $K^P_{m_1 \cdots m_n}$, for $n = 1, 2, 3, \cdots$, denote the corresponding nested sequence of intervals in K. If E_0 denotes the set of P for which $\bigcap_{n=1}^{\infty} K^P_{m_1 \cdots m_n} = \varnothing$, then $f(E_0) = 0$. If P is not in E_0, the corresponding intersection will consist of a single point or an interval. It is easily seen that if $P \neq P'$, then $\bigcap_{n=1}^{\infty} K^P_{m_1 \cdots m_n}$ and $\bigcap_{n=1}^{\infty} K^{P'}_{m_1' \cdots m_n'}$ are disjoint; therefore, at most a countable number of points P can exist for which the intersection is an interval. The mapping α is then defined to associate with P not in E_0 the unique point Q in $\bigcap_{n=1}^{\infty} K^P_{m_1 \cdots m_n}$ or the interval that this intersection forms. The definition of β is similar: For a fixed point Q in K, there is a unique sequence $K_{m_1 \cdots m_n}$ of intervals which is nested and such that each interval contains Q. Let $J^Q_{m_1 \cdots m_n}$, for $n = 1, 2, 3, \cdots$, denote the corresponding sequence of squares. Radon showed that $\overline{E}_0$, the set of Q for which $\bigcap_{n=1}^{\infty} J^Q_{m_1 \cdots m_n} = \varnothing$, has measure zero. When Q is not in $\overline{E}_0$ the intersection contains a single point, P, and $\beta(Q)$ is taken to be P.

Once Theorem 3 is established, the proof of Theorem 2 is relatively easy. Corresponding to the given set function g of Theorem 2 is the set function G defined by

(6) $$G(E') = g(E) + \sum_i [m(E' \cap A_i)/m(A_i)]g(\{P_i\}),$$

where E' is any measurable subset of K and E is the corrresponding f-measurable set defined in (iv) of Theorem 3. It is easily seen that G is additive and absolutely continuous. (If $m(E') = 0$, then $G(E') = 0$.) Therefore, by Lebesgue's Theorem 1, there exists a summable function Φ such that

(7) $$G(E') = \int_{E'} \Phi(Q)dQ.$$

By virtue of the way G is defined, it can be assumed without loss of generality that the function takes the constant value $g(\{P_i\})/m(A_i)$ on the interval A_i for $i = 1, 2, 3, \cdots$. Radon then defined a function Ψ

on $J - E_0$ by $\Psi(P) = \Phi(\alpha(P))$. The properties of the correspondence $E \leftrightarrow E'$ defined in Theorem 3 together with (6) and (7) show that Ψ is f-summable and that for every f-measurable set E,

$$g(E) = \int_E \Psi(P) df.$$

It should be clear that Radon's contribution to integration theory depended heavily upon the ideas of others, particularly Lebesgue. Both in his definition of $\int_J F df$ and in his proof of Theorem 2, Radon's originality, in essence, was the recognition of the fact that much of what had already been said in the context of Lebesgue measure could be said with reference to additive set functions. He perceived that Lebesgue's ideas would remain viable in a much more general setting, and his vision has been borne out by the diverse theories and applications of abstract measure and integration theory that are now commonplace in mathematics.

REFERENCE
MATERIAL

Dini's Theorem
on the Differentiability of
Continuous Functions

As NOTED in 2.3, Dini [1878: 188] proved the following theorem:

> *If f is continuous on [a,b] and if f(x)+Ax+B possesses at most a finite number of maxima and minima on [a,b] for all values of A except possibly a finite number, then f'(x) exists (as a finite number) on a dense subset of [a,b].*

Dini's method of proof will be amply illustrated if it is specialized to the case in which there are no exceptional values of A. The general proof requires a few modifications which are not particularly interesting.

It will be necessary to recall some definitions given in 2.3. For $I = [p,q]$ a subinterval of $[a,b]$, Dini considered the function

$$g(x;p,q) = f(x) - f(p) - \frac{f(q) - f(p)}{q - p} (x - p),$$

which, of course, is of the form $f(x)+Ax+B$. Also, let l_I and L_I denote the greatest lower bound and the least upper bound, respectively, of the quotient

$$\frac{f(x + h) - f(x)}{h},$$

for x and $x+h$ in I and $h \neq 0$.

Let $[c,d]$ denote an arbitrary subinterval of $[a,b]$. Dini proved the theorem by establishing the existence of an x_0 in (c,d) such that

$f'(x_0)$ exists. The hypothesis of the theorem guarantees the existence of subintervals I of $[c,d]$ on which f is monotonic, so either l_I or L_I must be finite. Assume without loss of generality therefore that the l corresponding to $[c,d]$ is itself finite. The heart of the proof is contained in the following lemma:

> Given any interval I contained in $[c,d]$ and any $\epsilon > 0$, there exists an interval J contained in the interior of I such that for all x and $x+h$ in J with $h \neq 0$,

$$(1) \qquad l_I \leq \frac{f(x+h) - f(x)}{h} \leq l_I + \epsilon.$$

The proof of the lemma is as follows: By definition of l_I there exists an r and s in I, with $r < s$, such that

$$(2) \qquad \frac{f(s) - f(r)}{s - r} < l_I + \epsilon.$$

Since $g(x;r,s)$ possesses a finite number of maxima and minima by hypothesis and since g takes the value 0 at r and s, it follows that there exists an interval J, contained in $[r,s]$, on which g is decreasing. In terms of f, this means that

$$(3) \qquad l_I \leq \frac{f(x+h) - f(x)}{h} \leq \frac{f(s) - f(r)}{s - r}$$

for all x and $x+h$ in J with $h \neq 0$. In combination, (2) and (3) imply (1).

The lemma is now applied to prove the existence of an x_0 in (c,d) for which $f'(x_0)$ exists: Let $I_1 = [c,d]$ and $\epsilon = 1/2$. Then there exists an interval I_2 contained in the interior of I_1 such that

$$l_1 \leq \frac{f(x+h) - f(x)}{h} \leq l_1 + 1/2$$

for all x and $x+h$ in I_2, where $l_1 = l_{I_1}$. Next apply the lemma to I_2 with $\epsilon = 1/2^2$ to obtain an interval I_3 in the interior of I_2 such that

$$l_2 \leq \frac{f(x+h) - f(x)}{h} \leq l_2 + 1/2^2$$

for all x and $x+h$ in I_3. Proceeding in this manner, we obtain a nested sequence of intervals $I_1 \supset I_2 \supset I_3 \supset \cdots \supset I_n \supset \cdots$ such that

$$(4) \qquad l_n \leq \frac{f(x+h) - f(x)}{h} \leq l_n + 1/2^n$$

for all x and $x+h$ in I_{n+1}, where I_{n+1} is contained in the interior of I_n. The sequence l_n is clearly increasing, and

$$l_{n+1} \leq l_1 + 1/2 + 1/2^2 + \cdots + 1/2^n < l_1 + 1,$$

so $l_0 = \lim_{n \to \infty} l_n$ is finite. Also, there exists an x_0 which is interior to all the I_n, and, by virtue of (4), $f'(x_0) = l_0$.

Glossary

ABSOLUTELY CONTINUOUS FUNCTION. A function $y = F(x)$ is absolutely continuous if for every $\epsilon > 0$ there exists a $\delta > 0$ such that for any sequence of intervals $[a_n, b_n]$, for $n = 1, 2, 3, \cdots$ with $\sum_{n=1}^{\infty} (b_n - a_n) < \delta$, it follows that $\sum_{n=1}^{\infty} |F(b_n) - F(a_n)| < \epsilon$.

ALMOST EVERYWHERE (A.E.). A property is said to hold amost everywhere, or a.e., with respect to a set E if it holds at all points of E with the possible exception of a set of points of E of Lebesgue measure zero.

BOREL SET. A set that is measurable in Borel's sense (see Section 4.2). More precisely and in modern terminology, one says that the Borel sets in $[a,b]$ are the members of the smallest σ-algebra in $[a,b]$ which contains all intervals. A σ-algebra in $[a,b]$ is a class Σ of subsets of $[a,b]$ such that: (i) $[a,b]$ is in Σ; (ii) if E is in Σ, then $[a,b] - E$ is in Σ; (iii) if E_n is in Σ for $n = 1, 2, 3, \cdots$, then $\bigcup_{n=1}^{\infty} E_n$ is in Σ.

BOUNDARY OF A SET. The set ∂E of boundary points of a set E is called its boundary. A point is a boundary point of E if every open set containing it contains points of E and points not in E. A set is open if its complement—the set of points not belonging to it—is a CLOSED SET.

BOUNDED VARIATION. A function f is said to be of bounded variation

200

on $[a,b]$ if there exists a constant M such that $\sum_{i=1}^{n} | f(x_i) - f(x_{i-1}) | \leq M$ for every partition $a = x_0 < x_1 < \cdots < x_n = b$ of $[a,b]$.

CHARACTERISTIC FUNCTION. A characteristic function is a function of the form $f(x) = \chi_E(x)$ for some set E, where $\chi_E(x) = 1$ for x in E and $\chi_E(x) = 0$ for x not in E.

CLOSED SET. A set is closed if it contains all its limit points. See Section 2.1.

COMPACT SET. A set E with the property that if $E \subset \bigcup_{\alpha} O_\alpha$ and the O_α are open sets, then there exists a finite number of the O_α whose union contains E. See HEINE-BOREL THEOREM.

DENSE SET. A set E is dense in $[a,b]$ if every subinterval of $[a,b]$ contains points of E.

DISCRETE SET. Term used by A. Harnack for a set with zero outer content.

FIRST SPECIES FUNCTION. Term used by Dini to denote a continuous function f for which the quotient $\dfrac{f(x+h) - f(x)}{h}$ is either bounded above or below for all x and $x+h$, $h \neq 0$, in the interval under consideration. These functions are always of BOUNDED VARIATION.

FIRST SPECIES SET. A term introduced by Cantor for any set E such that $E^{(n)} = \varnothing$ for some integer n, where the derived sets $E^{(n)}$ are defined by $E^{(1)} = E'$—the set of limit points of E—and $E^{(n)} = (E^{(n-1)})'$.

FUBINI'S THEOREM. A generic term used to refer to any theorem asserting the identity of a double integral with iterated single integrals.

FUNDAMENTAL THEOREM I. A generic term used to refer to any theorem asserting that $\left(\int_a^x f \right)' = f(x)$ for "certain" values of x. See Section 1.2 for Cauchy's version of this theorem.

FUNDAMENTAL THEOREM II. A generic term used to refer to any theorem asserting that $\int_a^b f' = f(b) - f(a)$. See Section 1.2 for Cauchy's version of this theorem.

FUNDAMENTAL THEOREM III. A generic term used to refer to any theorem asserting that if $F'(x)$—or some generalized derivative of F at x—vanishes on a certain set of x, then F is a constant function. See Section 1.2 for Cauchy's version.

HEINE-BOREL THEOREM. In its most general form, the theorem states that a CLOSED and bounded subset of n-dimensional Euclidean space is COMPACT. Borel proved this result for closed intervals. His proof used a method somewhat similar to that employed by Heine to prove that a function continuous on a closed interval is uniformly continuous there, but Heine's proof apparently goes back to Weierstrass. The term "Heine-Borel Theorem" for Borel's theorem was introduced by A. Schoenflies in his report on the theory of sets [1900] discussed in Section 4.3.

INTEGRABLE FUNCTION. A bounded function that is integrable in Riemann's sense, unless a different meaning is explicitly mentioned.

MEASURABLE FUNCTION. A function f such that for every real number a, the set of x for which $f(x) > a$ is Lebesgue-measurable. See Section 5.1.

NORM OF A PARTITION. The norm of the partition P of $[a,b]$ defined by $a = x_0 < x_1 < \cdots < x_n = b$ is the maximum of the differences $x_i - x_{i-1}$, $i = 1, 2, \cdots, n$, and is denoted by $\|P\|$. The norm of a partition of $(-\infty, \infty)$ is defined in a similar manner.

NOWHERE DENSE SET. A set E is nowhere dense in $[a,b]$ if given any subinterval I of $[a,b]$ there exists an interval J contained in I which contains no points of E.

OSCILLATION OF A FUNCTION AT A POINT. If f is defined in an open interval about the point x, the oscillation of f at x is the number $\omega_f(x) = \lim_{\delta \to 0} [M(\delta) - m(\delta)]$, where $M(\delta)$ and $m(\delta)$ denote the least upper bound and greatest lower bound, respectively, of $f(t)$ for $|t - x| < \delta$.

REDUCIBLE SET. A set E with the property that $E^{(\gamma)} = \varnothing$ for some ordinal number γ less than Ω. See Section 3.3.

RIEMANN'S INTEGRABILITY CONDITION (R_1). If $f(x)$ is defined and bounded on $[a,b]$, it satisfies Riemann's condition (R_1) if

$$\lim_{\|P\| \to 0} (D_1\delta_1 + D_2\delta_2 + \cdots + D_n\delta_n) = 0,$$

where P denotes a partition $a = x_0 < x_1 < \cdots < x_n = b$ of $[a,b]$, $\delta_i = x_i - x_{i-1}$, and D_i is the oscillation of $f(x)$ on $[x_{i-1}, x_i]$, i.e., $D_i = M_i - m_i$, where

M_i and m_i denote, respectively, the least upper bound and greatest lower bound of $f(x)$ on $[x_{i-1}, x_i]$. Later (R_1) was transformed into the condition that the upper and lower integrals of f be equal: $\overline{\int_a^b} f(x)dx = \underline{\int_a^b} f(x)dx$.

RIEMANN'S INTEGRABILITY CONDITION (R_2). Let the function f and partition P be as in RIEMANN'S INTEGRABILITY CONDITION (R_1). Then f satisfies (R_2) if for every $\sigma > 0$ and $\epsilon > 0$ there exists a $\delta > 0$ such that $s(P, \sigma) < \epsilon$ for any partition P with NORM $\|P\| < \delta$, where $s(P, \sigma)$ denotes the sum of those δ_i such that $D_i > \sigma$ in the notation of (R_1). Later (R_2) was expressed in the following form: For every $\sigma > 0$, $c_e(S_\sigma) = 0$, where S_σ is the set of all x such that the OSCILLATION of f at x is greater than σ.

SUMMABLE FUNCTION. A function is summable on $[a,b]$ if it is integrable in Lebesgue's sense. Any bounded MEASURABLE FUNCTION is summable. An unbounded measurable function f is summable on $[a,b]$ provided $\sum_{-\infty}^{\infty} |a_i| m(e_i) < \infty$, where $\cdots < a_{-2} < a_{-1} < a_0 < a_1 < a_2 < \cdots$, $a_i - a_{i-1}$ is bounded in i, and e_i is the set of all x in $[a,b]$ such that $a_i \leq f(x) < a_{i+1}$.

UNIFORMLY BOUNDED SERIES. A series of functions $f(x) = \sum_{n=1}^{\infty} u_n(x)$ is uniformly bounded on a set E if a number M exists such that $|S_n(x)| \leq M$ for all x in E and all n, where $S_n(x) = \sum_{k=1}^{n} u_k(x)$.

UNIFORMLY CONVERGENT SERIES. A series of functions $f(x) = \sum_{n=1}^{\infty} u_n(x)$ converges uniformly on a set E if for any $\epsilon > 0$ an integer N exists such that $|R_n(x)| < \epsilon$ for all $n \geq N$ and for all x in E, where $R_n(x) = \sum_{k=n+1}^{\infty} u_k(x)$.

SPECIAL SYMBOLS

χ_E Characteristic function of the set E

$\|P\|$ Norm of the partition P

$\omega_f(x)$ Oscillation of a function at a point

∂E Boundary of a set E

$\varnothing$ The null or empty set

List of Abbreviations

Abh. Gesch. Math.	Abhandlungen zur Geschichte der Mathematik
Acta Math.	Acta Mathematica
Amer. Jl. Math.	American Journal of Mathematics
Amer. Math. Soc. Trans.	Transactions of the American Mathematical Society
Ann. Inst. Fourier	Annales de l'Institut Fourier
Ann. Mat.	Annali di Mathematica pura ed applicata
Ann. Math.	Annals of Mathematics
Arch. Hist. Exact Sci.	Archive for History of Exact Science
Arch. Math. Phys.	Archiv der Mathematik und Physik
Berlin Ak. Abh.	Abhandlungen der Königlichen Akademie der Wissenschaften zu Berlin
Berlin Ak. Monatsber.	Monatsberichte der Königlich Preusschen Akademie der Wissenschaften zu Berlin
Berlin Ak. Sber.	Sitzungsberichte der Königlich Preussischen Akademie der Wissenschaften zu Berlin
Bologna Acc. Sci. Mem.	Memorie della [R.] Accademia delle Scienze dell'Instituto di Bologna
Brit. Ass. Rep.	Report of the Third Meeting of the British Association for the Advancement of Science. Held at Cambridge in 1833.
Brux., Ac. Bull.	Bulletins de l'Académie Royale des Sciences, des Lettres et des Beaux-Arts de Belgique
Brux., Ac. Mém.	Mémoires de l'Académie Royale des Sciences, des Lettres et des Beaux-Arts de Belgique
Brux., Ac. Sci. Bull.	Académie Royale de Belgique. Bulletin de la classe des sciences. Bruxelles
Brux., Mém. Couronn.	Mémoires Couronnés et Mémoires des Savants

	Étrangers publiés par l'Académie Royale des Sciences, des Lettres et des Beaux-Arts de Belgique
Brux. Soc. Sci. Ann.	Annales de la Société Scientifique de Bruxelles
Bull. Sci. Math.	Bulletin des Sciences Mathématiques
Cambridge Phil. Soc. Proc.	Proceedings of the Cambridge Philosophical Society
Cambridge Phil. Soc. Trans.	Transactions of the Cambridge Philosophical Society
Catania Acc. Gioen. Bull.	Bullettino Mensile della Accademia Gioenia di Scienze Naturali in Catania [New Series]
Crelle, Jl. Math.	Journal für die reine und angewandte Mathematik
Deutsch. Math. Ver. Jber.	Jahresbericht der Deutschen Mathematiker-Vereinigung
Fortschr. Math.	Jahrbuch über die Fortschritte der Mathematik
Giorn. Mat.	Giornale di Matematiche
Göttingen Nachr.	Nachrichten von der Königl. Gesellschaft der Wissenschaften zu Göttingen. Mathematisch-physikalische Klasse
Helsingfors, Acta	Acta Societatis Scientiarum Fennicae
Irish Ac. Trans.	The Transactions of the Royal Irish Academy
Liouville, Jl. Math.	Journal de Mathématiques pures et appliquées . . .
London Math. Soc. Proc.	Proceedings of the London Mathematical Society
Manchester Lit. Phil. Soc. Mem.	Memoirs of the Literary and Philosophical Society of Manchester
Math. Ann.	Mathematische Annalen
Milano, Ist. Lomb. Rend.	Reale Istituto Lombardo di Scienze e Lettere. Rendiconti
Modena Mem. Soc. Ital.	Memorie di Matematiche e di Fisica della Società Italiana della Scienze, Modena
Monhfte Math. Phys., Wien	Monatshefte für Mathematik und Physik
München, Ak. Abh.	Abhandlungen der mathematisch-physikalischen Classe der k. b. Akademie der Wissenschaften
München, Ak. Sber.	Sitzungsberichte der mathematisch-physikalischen Classe der k. b. Akademie der Wissenschaften zu München
Palermo Circ. Mat. Rend.	Rendiconti del Circolo Matematico di Palermo
Paris, Ac. Sci. C.R.	Comptes Rendus hebdomadaires des Séances de l'Académie des Sciences
Paris, Ac. Sci. Mém.	Mémoires de l'Académie des Sciences
Paris, École Norm. Ann.	Annales Scientifiques de l'École Normale Supérieure
Paris, École Polyt. Jl.	Journal de l'École Polytechnique
Paris, Soc. Math. Bull.	Bulletin de la Société Mathématique de France
Phil. Trans.	Philosophical Transactions of the Royal Society of London
Pisa, Univ. Toscane Ann.	Annali delle Università Toscane

Quart. Jl. Math.	The Quarterly Journal of Pure and Applied Mathematics
Repert. der Phys.	Repertorium der Physik
Roma, R. Acc. Lincei Rend.	Atti della R. Accademia dei Lincei. Rendiconti
Roma, R. Acc. Lincei Trans.	Atti della R. Accademia dei Lincei. Transunti
Roy. Soc. Proc.	Proceedings of the Royal Society of London
Schweiz. Natf. Ges. Verh.	Verhandlungen der Schweizerischen Naturforschenden Gesellschaft
Stockh., Öfvers.	Öfversigt af Kongl. Vetenskaps-Akademiens Förhandlingar
Torino Acc. Sci. Atti	Atti della R. Accademia delle Scienze di Torino [Classe di Scienze Fisiche, Matematiche e Naturali]
Toulouse Ann.	Annales de la Faculté des Sciences de Toulouse, pour les Sciences Mathématiques et les Sciences Physiques
Wien, Ak. Sber.	Sitzungsberichte der kaiserlichen Akademie der Wissenschaften.Mathematisch-Naturwissenschaftliche Classe
Zeitschr. Math. Phys.	Zeitschrift für Mathematik und Physik

Bibliography

Abel, N. H., 1826. "Recherches sur le série $1+(m/1)x+(m(m-1)/1\cdot2)x^2$ $+(m(m-1)(m-2)/1\cdot2\cdot3)x^3+\cdots$." *Crelle, Jl. Math.*, *1*, 221–24.

Ampère, A. M., 1806. "Recherches sur quelques points de la théorie des fonctions derivées qui conduissent à une nouvelle démonstration de la série de Taylor." *Paris École Polyt. Jl.*, *13*, 148–81.

Appell, P., 1882a. "Développements en série d'une fonction holomorphe dans une aire limitée par des arcs de cercle." *Paris, Ac. Sci. C. R.*, *94*, 1238–40.

Appell, P., 1882b. "Développements en série dans une aire limitée par des arcs de cercle." *Acta Math.*, *1*, 145–52.

Arzelà, Cesare, 1885a. "Un teorema intorno alle serie di funzioni." *Roma, R. Acc. Lincei Rend.*, (*4*) *1*, 262–67.

Arzelà, Cesare, 1885b. "Sulla integrabilità di una serie di funzioni." *Roma, R. Acc. Lincei Rend.*, (*4*) *1*, 321–26.

Arzelà, Cesare, 1885c. "Sulla integrazione per serie." *Roma R. Acc. Lincei Rend.*, (*4*) *1*, 532–37, 566–69.

Arzelà, Cesare, 1891. "Sugli integrali doppi." *Bologna Acc. Sci. Mem.*, (*5*) *2*, 133–47.

Arzelà, Cesare, 1899–1900. "Sulle serie di funzioni." *Bologna Acc. Sci. Mem.*, (*5*) *8*, 131–86, 701–44.

Ascoli, G., 1873. "Ueber trigonometrische Reihen." *Math. Ann.*, *6*, 231–40.

Ascoli, G., 1875. "Sul concetto di integrale definito." *Roma, R. Acc. Lincei*, (*2*) *2*, 863–72.

Ascoli, G., 1883. "Il concetto di lunghezza di curva è indipendente da quello di derivata." *Milano Ist. Lomb. Rend.*, (*2*) *16*, 851–53.

Ascoli, G., 1884. "Il concetto di lunghezza di linea non è soltanto indipendente dal concetto di derivata ma anche da quello di continuità." *Milano Ist. Lomb. Rend.*, (*2*) *17*, 567–72.

Baire, R., 1898. "Sur les fonctions discontinues qui rattachent aux fonctions continues." *Paris, Ac. Sci., C. R., 126,* 1621–23.

Baire, R., 1899. "Sur les fonctions de variables réelles." *Ann. Mat.,* (*3*) *3*, 1–122.

Baire, R., 1905. *Leçons sur les fonctions discontinues.* Paris, Gauthier-Villars.

Bendixson, Ivar, 1883. "Quelques théorèmes de la théorie des ensembles." *Acta Math., 2,* 415–29.

Bernkopf, Michael, 1966. "The Development of Function Spaces with Particular Reference to Their Origins in Integral Equation Theory." *Arch. Hist. Exact Sci., 3,* 1–96.

Bertrand, Joseph, 1864–70. *Traité de calcul différentiel et de calcul intégral.* 2 vols. Paris, Gauthier-Villars.

Bettazzi, Rodolfo, 1884. "Sui concetti di derivazione e d'integrazione delle funzioni di piu variabili reali." *Giorn. Mat., 22,* 133–67.

Bolzano, B., 1930. *Functionenlehre. Schriften,* Vol. 1. Prague, Karl Petr.

Boole, G., 1848. "On the Analysis of Discontinuous Functions." *Irish Ac. Trans., 21,* 124–39.

Borel, Émile, 1895. "Sur quelques points de la théorie des fonctions." *Paris, École Norm. Ann.,* (*3*) *12*, 9–55. Also published separately as a doctoral thesis (Paris, 1894).

Borel, Émile, 1898. *Leçons sur la théorie des fonctions.* Paris, Gauthier-Villars.

Borel, Émile, 1905. *Leçons sur les fonctions de variables réelles.* Paris, Gauthier-Villars.

Borel, Émile, 1912. *Notice sur les travaux scientifiques de M. Émile Borel,* 2nd ed. Paris, Gauthier-Villars.

Borel, Émile, 1940. *Selecta: Jubilé scientifique de M. Émile Borel.* Paris, Gauthier-Villars.

Borel, Émile, and Jules Drach, 1895. *Introduction a l'étude de la théorie des nombres et de l'algèbre supériure.* Paris, Nony.

Bourbaki, N., 1960. *Éléments d'histoire des mathématiques.* Paris, Hermann.

Boussinesq, J., 1887–90. *Cours d'analyse infinitésimale.* . . . 4 vols. Paris, Gauthier-Villars.

Boyer, C. B., 1959. *The History of the Calculus and its Conceptual Development,* 2nd ed. New York, Dover. Reprint of *The Concepts of the Calculus, A Critical and Historical Discussion of the Derivative and the Integral* (New York, Hafner, 1949).

Brodén, T., 1896. "Ueber das Weierstrass-Cantor'sche Condensationsverfahren." *Stockh. Öfvers.*, *53*, 583–602.

Brodén, T., 1900. "Derivirbare Funktionen mit überall dichten Maxima und Minima." *Stockh. Öfvers.*, *57*, 423–41, 743–61.

Burkill, J. C., 1942–44. "Henri Lebesgue." *Obituary Notices of Fellows of the Royal Society*, *4*, 483–90.

Cajori, Florian, 1919. *A History of Mathematics*, 2nd ed. New York, Macmillan.

Cantor, Georg, 1870a. "Ueber einen die trigonometrischen Reihen betreffenden Lehrsatz." *Crelle, Jl. Math.*, *72*, 130–38.

Cantor, Georg, 1870b. "Beweis, dass eine für jeden reellen Werth von x durch eine trigonometrische Reihe gegebene Function $f(x)$ sich nur auf eine einzige Weise in dieser Form darstellen lässt," Pt 1. *Crelle, Jl. Math.*, *72*, 139–42.

Cantor, Georg, 1871a. "Beweis, dass eine für jeden reellen Werth von x durch eine trigonometrische Reihe gegebene Function $f(x)$ sich nur auf eine einzige Weise in dieser Form darstellen lässt," Pt 2. *Crelle, Jl. Math.*, *73*, 294–96.

Cantor, Georg, 1871b. Review of [Hankel 1870]. *Literarisches Centralblatt für Deutschland*, (February), 150–51.

Cantor, Georg, 1872. "Ueber die Ausdehnung eines Satzes aus der Theorie der trigonometrischen Reihen." *Math. Ann.*, *5*, 92–101.

Cantor, Georg, 1874. "Über eine Eigenschaft des Inbegriffs aller reellen algebraischen Zahlen." *Crelle, Jl. Math.*, *77*, 258–62.

Cantor, Georg, 1879. "Ueber unendliche, lineare Punktmannichfaltigkeiten," Pt 1. *Math. Ann.*, *15*, 1–7.

Cantor, Georg, 1880. "Ueber unendliche, lineare Punktmannichfaltigkeiten," Pt 2. *Math. Ann.*, *17*, 355–58.

Cantor, Georg, 1882a. "Über ein neues und allgemeines Kondensationsprinzip der Singularitäten von Funktionen." *Math. Ann.*, *19*, 588–94.

Cantor, Georg, 1882b. "Ueber unendliche, lineare Punktmannichfaltigkeiten," Pt 3. *Math. Ann.*, *20*, 113–21.

Cantor, Georg, 1883. "Ueber unendliche, lineare Punktmannichfaltigkeiten," Pt 4. *Math. Ann.*, *21*, 51–58, 545–91.

Cantor, Georg, 1884a. "Ueber unendliche, lineare Punktmannichfaltigkeiten," Pt 5. *Math. Ann.*, *23*, 453–88.

Cantor, Georg, 1884b. "De la puissance des ensembles parfaits de points: Extrait d'une lettre adressé à l'éditeur." *Acta Math.*, *4*, 381–92.

Carathéodory, C., 1914. "Über das lineare Mass von Punktmengen—eine Verallgemeinerung des Längenbegriffes." *Göttingen Nachr.*, 404–26.

Cauchy, Augustin, 1821. *Cours d'Analyse de l'École Royale Polytechnique.* Paris, deBure. Page references are from *Oeuvres complètes*, *(2) 3* (1897).

Cauchy, Augustin, 1823. *Résumé des Leçons données à l'École Royale Polytechnique sur le calcul infinitésimale.* Paris, deBure. Page references are from *Oeuvres complètes*, *(2) 4* (1899).

Cauchy, Augustin, 1827a. "Mémoire sur les intégrales définies." *Paris, Ac. Sci. Mem., 1.* Page references are from *Oeuvres complètes*, *(1) 1* (1882), 319–506.

Cauchy, Augustin, 1827b. "Mémoire sur les développements des fonctions en séries périodiques." *Paris, Ac. Sci. Mem., 6,* 603ff. Page references are from *Oeuvres complètes*, *(1) 2* (1908), 12–19.

Cauchy, Augustin, 1829. *Leçons sur le calcul differentiel.* Paris, deBure. Page references are from *Oeuvres complètes*, *(2) 4* (1899), 263–609.

Cauchy, Augustin, 1841. "Mémoire sur le rapport différentiel de deux grandeurs qui varient simultanément." *Exercices d'analyse et de physique mathématique,* Vol. 2, pp. 188–229. Page references are from *Oeuvres complètes (2) 12* (1916), 214–62.

Cauchy, Augustin, 1849. "Mémoire sur les fonctions discontinues." *Paris, Ac. Sci. C. R., 28,* 277ff. Page references are from *Oeuvres complètes, (1) 11* (1899), Extract 435, pp. 120–26.

Cauchy, Augustin, 1853. "Note sur les séries convergentes dont les divers terms sont des fonctions continues d'une variable réelle ou imaginaire, entre des limites données." *Paris, Ac. Sci. C.R., 36,* 454–59. Page references are from *Oeuvres complètes, (1) 12* (1900), Extract 518, pp. 30–36.

Cauchy, Augustin, 1882–1958. *Oeuvres complètes d'Augustin Cauchy.* 26 vols. Paris, Gauthier-Villars.

Cellérier, C., 1890. "Note sur les principes fondamentaux de l'analyse." *Bull. Sci. Math., (2) 14,* 142–60.

Daniell, P. J., 1917–18. "A General Form of the Integral." *Ann. Math., (2) 19,* 279–94.

Darboux, Gaston, 1875. "Mémoire sur la théorie des fonctions discontinues." *Paris, École Norm. Ann., (2) 4,* 57–112.

Darboux, Gaston, 1879. "Addition au mémoire sur les fonctions discontinues." *Paris, École Norm. Ann., (2) 8,* 195–202.

Darboux, Gaston, 1881. *Notice sur les travaux scientifiques de M. Gaston Darboux,* 1st ed. Paris, Gauthier-Villars. 2nd ed., 1884.

de la Vallée-Poussin, C., 1892. "Sur la convergence des intégrales défines." *Liouville, Jl. Math.*, (*4*) *8*, 421–67.

de la Vallée-Poussin, C., 1893. Sur quelques applications de l'intégrale de Poisson." *Brux., Soc. Sci. Ann.*, *17*, 18–34.

de la Vallée-Poussin, C., 1899. "Réduction des intégrales multiples généralizées." *Liouville, Jl. Math.*, (*5*) *5*, 191–204.

de la Vallée-Poussin, C., 1903. *Cours d'analyse infinitésimale*, Vol. 1. Louvain and Paris.

de la Vallée-Poussin, C., 1910. "Réduction des intégrales doubles de Lebesgue: Application à la définition des fonctions analytiques." *Brux., Ac. Sci. Bull.*, 768–98.

Denjoy, Arnaud, 1912a. "Une extension de l'intégrale de M. Lebesgue." *Paris, Ac. Sci. C. R.*, *154*, 859–62.

Denjoy, Arnaud, 1912b. "Calcul de la primitive de la fonction dérivée la plus générale." *Paris, Ac. Sci. C. R.*, *154*, 1075–78.

Denjoy, Arnaud, 1916. "Mémoire sur la totalisation des nombres dérivées non sommables." *Paris, École Norm. Ann.*, (*3*) *33*, 127–222.

Dickstein, S., 1899. "Zur Geschichte der Prinzipien der Infinitesmalrechnung: Die Kritiker der 'Theorie des Fonctions Analytiques' von Lagrange." *Abh. Gesch. Math.*, *9*, 65–79.

Dini, Ulisse, 1874. "La serie di Fourier." *Pisa, Univ. Toscane Ann.*, *14*, 161–76.

Dini, Ulisse, 1877a. "Sopra una classe di funzioni finite e continue, che non hanno mai una derivata." *Roma, R. Acc. Lincei Trans.*, (*3*) *1*, 70–72.

Dini, Ulisse, 1877b. "Su alcune funzioni che in tutto un intervallo non hanno mai derivata." *Ann. Mat.*, (*2*), *8*, 121–37.

Dini, Ulisse, 1877c. "Sulle funzioni finite continue di variabili reali che non hanno mai derivata." *Roma, R. Acc. Lincei Trans.*, (*3*) *1*, 130–33.

Dini, Ulisse, 1878. *Fondamenti per la teorica della funzioni di variabili reali*. Pisa. Translated and supplemented by Jacob Lüroth and Adolf Schepp as *Grundlagen für eine Theorie der Functionen einer veränderlichen reellen Grösse* (Leipzig, B. G. Teubner, 1892).

Dini, Ulisse, 1880. *Serie di Fourier e altre rappresentazioni analitiche delle funzioni di una variabile reale*. Pisa, Eip. T. Nistri.

Dirichlet, P. G. Lejeune, 1829. "Sur la convergence des séries trigonométriques qui servent a représenter une fonction arbitraire entre des limites données." *Crelle, Jl. Math.*, *4*, 157–69.

Dirichlet, P. G. Lejeune, 1837. "Ueber die Darstellung ganz willkürlicher Functionen durch Sinus- und Cosinusreihen." *Repert. der Phys.*, *1*, 152–74.

Dirichlet, P. G. Lejeune, 1904. *G. Lejeune Dirichlet's Vorlesungen über die Lehre von den einfachen und mehrfachen bestimmten Integralen.* Edited by G. Arendt. Brunswick, Germany, F. Vieweg.

Dirksen, E. H., 1833. "Über die Anwendung der Analysis auf die Rectification der Curven, die Quadratur der Flächen und die Cubatur der Körper." *Berlin Ak. Abh.*, 123–68.

Drach, Jules, 1895. "Sur l'application aux équations différentielles de méthodes analogues à celles de Galois." *Paris, Ac. Sci. C.R., 120,* 73–76.

Drach, Jules, 1898. "Essai sur une theorie générale de l'intégration et sur la classification des transcendentes." *Paris, École Norm. Ann., (3) 15,* 243–384.

du Bois-Reymond, Paul, 1870. *Antrittsprogramm, enthaltend neue Lehrsätze über die Summen unendlicher Reihen zur Übernahme der ordentlichen Professur für Mathematik an der Universität Freiburg in Baden.* Berlin.

du Bois-Reymond, Paul, 1873. "Ueber die Fourier'schen Reihen." *Göttingen Nachr.*, 571–82.

du Bois-Reymond, Paul, 1875a. "Versuch einer Classification der willkürlichen Functionen reeller Argumente nach ihren Aenderungen in den kleinsten Intervallen." *Crelle, Jl. Math., 79,* 21–37.

du Bois-Reymond, Paul, 1875b. "Beweis, dass die Coeffizienten der trigonometrischen Reihe $f(x) = \sum_{p=0}^{\infty} (a_p \cos px + b_p \sin px)$ die Werthe

$$a_0 = \frac{1}{2\pi} \int_{-\pi}^{+\pi} d\alpha f(\alpha), \, a_p = \frac{1}{\pi} \int_{-\pi}^{+\pi} d\alpha f(\alpha) \cos p\alpha, \, b_p = \frac{1}{\pi} \int_{-\pi}^{+\pi} d\alpha f(\alpha) \sin p\alpha$$

haben, jedesmal wenn diese Integrale endlich und bestimmt sind. (Mit. einem Anhang: Über den Fundamantalsatz der Integralrechnung.)" *München, Ak. Abh., 12$_1$,* 117–66.

du Bois-Reymond, Paul, 1875c. Review of [Thomae 1875]. *Zeitschr. Math. Phys., 20,* 121–29.

du Bois-Reymond, Paul, 1876. "Untersuchungen über die Convergenz und Divergenz der Fourier'schen Darstellungsformeln." *München, Ak. Abh., 12$_2$,* 1–102.

du Bois-Reymond, Paul, 1879. "Erläuterungen zu den Anfangsgründen der Variationrechnung." *Math. Ann., 15,* 282–315, 564–76.

du Bois-Reymond, Paul, 1880a. *Zur Geschichte der trigonometrischen Reihen: Eine Entgegnung.* Tübingen, H. Laupp.

du Bois-Reymond, Paul, 1880b. "Der Beweis des Fundamentalsatzes der Integralrechnung: $\int_a^b F'(x)dx = F(b) - F(a)$." *Math. Ann., 16,* 115–28.

du Bois-Reymond, Paul, 1882. *Die allgemeine Functionentheorie.*

Tübingen, H. Laupp. Translated by G. Milhaud and A. Girot as *Théorie générale des fonctions* (Nice, 1887).

du Bois-Reymond, Paul, 1883a. "Ueber das Doppelintegral." *Crelle, Jl. Math.*, *94*, 273–90.

du Bois-Reymond, Paul, 1883b. "Ueber die Integration der trigonometrischen Reihe." *Math. Ann.*, *22*, 260–68.

du Bois-Reymond, Paul, 1885. "Ueber den Begriff der Länge einer Curve." *Acta Math.*, *6*, 167–68.

du Bois-Reymond, Paul, 1886. "Über die Integration der Reihen." *Berlin Ak. Sber.*, 359–71.

du Bois-Reymond, Paul, 1887. "Über den Convergenzgrad der variablen Reihen und den Stetigkeitsgrad der Functionen zweier Argumente." *Crelle, Jl. Math.*, *100*, 331–58.

Duhamel, J. M. C., 1847. *Cours d'analyse de l'École polytechnique*, 2nd ed. 2 vols. Paris, Bachelier.

Duhamel, J. M. C., 1856. *Éléments de calcul infinitésimale.* 2 vols. Paris, Mallet-Bachelier.

Duhamel, J. M. C., 1866. *Des méthodes dans les sciences de raisonnement*, Vol. 2. Paris, Gauthier-Villars.

Euler, Leonhard, 1748. *Introductio in analysin infinitorum.* 2 vols. Lausanne.

Fatou, P., 1906. "Séries trigonométriques et séries de Taylor." *Acta Math.*, *30*, 335–400.

Fejér, L., 1900. "Sur les fonctions bornées et intégrables." *Paris, Ac. Sci. C.R.*, *131*, 984–87.

Fejér, L., 1904. "Untersuchungen über Fouriersche Reihen." *Math. Ann.*, *58*, 51–69.

Fischer, Ernst, 1904. "Zwei neue Beweise für den 'Fundamentalsatz der Fourierschen Konstanten.' " *Monhfte Math., Phys., Wien, 15*, 69–92.

Fischer, Ernst, 1907a. "Sur la convergence en moyenne." *Paris, Ac. Sci. C.R.*, *144*, 1022–24.

Fischer, Ernst, 1907b. "Applications d'un théorème sur la convergence moyenne." *Paris, Ac. Sci. C.R.*, *144*, 1148–51.

Fourier, Joseph, 1822. *La théorie analytique de la chaleur.* Paris, Didot. Translated by Alexander Freeman as *The Analytical Theory of Heat*, Cambridge, 1878. Page references are from the English translation.

Fréchet, Maurice, 1915. "Sur l'intégrale d'une fonctionelle étendue à un ensemble abstrait." *Paris, Soc. Math. Bull.*, *43*, 248–65.

Fréchet, Maurice, 1965. "La vie et l'oeuvre d'Émile Borel." *L'Enseignement mathematique*, *(2) 11*, 1–94.

Freycinet, C., 1860. *De l'analyse infinitésimale: Étude sur la métaphysique du haut calcul. . . .* Paris, Mallet-Bachelier.

Fubini, G., 1907. "Sugli integrali multipli." *Roma, R. Acc. Lincei Rend.*, (5) 16₁, 608–14.

Galois, É., 1962. *Écrits et mémoires mathématiques d'Évariste Galois.* Edited by Robert Bourgne and J.-P. Azra. Paris, Gauthier-Villars.

Gauss, Karl F., 1863–1929. *Werke.* 12 vols. Leipzig, B. G. Teubner.

Gilbert, P., 1873a. "Mémoire sur l'existence de la dérivée dans les fonctions continues." *Brux., Mém. Couronn., 23*, No. 3.

Gilbert, P., 1873b. "Rectification au sujet d'un mémoire précédent." *Brux., Ac. Bull.*, (2) 35, 709–17.

Gilbert, P., 1892. *Cours d'analyse infinitésimale*, 4th ed. Paris, Gauthier-Villars.

Goursat, E., 1882. "Sur les fonctions uniformes présentant des lacunes." *Paris, Ac. Sci. C.R., 94*, 715–18.

Goursat, E., 1887. "Sur les fonctions à espaces lacunaires." *Bull. Sci. Math.*, (2) 11, 109–14.

Hadamard, J., 1898. *Leçons de géométrie élémentaire*, Vol. 1. Paris, A. Colin.

Hadamard, J., 1903. "Sur les opérations fonctionnelles." *Paris, Ac. Sci. C.R., 136*, 351–54.

Hahn, H., 1905. "Über den Fundamentalsatz der Integralrechnung." *Monhfte Math. Phys., Wien, 16*, 161–66.

Halphén, G., 1882. "Sur la série de Fourier." *Paris, Ac. Sci. C.R., 95*, 1217–19.

Hankel, H., 1870. *Untersuchungen über die unendlich oft oszillierenden und unstetigen Functionen.* Presented at the University of Tübingen, March 6, 1870. Page references are from the reprinting in *Math. Ann., 20* (1882), 63–112.

Hankel, H., 1871. "Grenze." *Allgemeine Encyklopädie der Wissenschaften und Künste*, Sec. I, Pt 90, pp. 189–211. Leipzig, F. A. Brockhaus.

Hardy, G. H., 1918. "Sir George Stokes and the Concept of Uniform Convergence." *Cambridge Phil. Soc. Proc., 19*, 148–56.

Hardy, G. H., 1943. "William Henry Young." *Obituary Notices of Fellows of the Royal Society, 4*, 307–23.

Harnack, Axel, 1880. "Über die trigonometrische Reihe und die Darstellung willkürlicher Functionen." *Math. Ann., 17*, 122–32.

Harnack, Axel, 1881. *Die Elemente der Differential- und Integralrechnung.* Leipzig, B. G. Teubner.

Harnack, Axel, 1882a. "Vereinfachung der Beweise in der Theorie der Fourier'schen Reihe." *Math. Ann., 19*, 235–79.

Harnack, Axel, 1882b. "Berichtigung dem Aufsatz: 'Ueber die Fourier'sche Reihe.'" *Math. Ann., 19*, 524–28.

Harnack, Axel, 1882c. "Théorie de la série de Fourier." *Bull. Sci. Math.*, (2), *6*, 242–60, 265–80, 282–300.

Harnack, Axel, 1883. "Anwendung der Fourier'schen Reihe auf die Theorie der Functionen einer complexen Veränderlichen." *Math. Ann.*, *21*, 305–26.

Harnack, Axel, 1884a. "Die allgemeinen Sätze über den Zusammenhang der Functionen einer reellen Variabelen mit ihren Ableitungen," Pt 1. *Math. Ann.*, *23*, 244–84.

Harnack, Axel, 1884b. "Notiz über die Abbildung einer stetigen linearen Mannigfaltigkeit auf eine unstetige." *Math. Ann.*, *23*, 285–88.

Harnack, Axel, 1884c. "Die allgemeinen Sätze über den Zusammenhang der Functionen einer reellen Variabelen mit ihren Ableitungen," Pt 2. *Math. Ann.*, *24*, 217–52.

Harnack, Axel, 1884–85. *Lehrbuch der Differential- und Integralrechnung.* 2 vols. Leipzig, B. G. Teubner. Translation, with supplementation, of [Serret 1868].

Harnack, Axel, 1885. "Ueber den Inhalt von Punktmengen." *Math. Ann.*, *25*, 241–50.

Harnack, Axel, 1886. "Bemerkung zur Theorie des Doppelintegrals." *Math. Ann.*, *26*, 566–68.

Hausdorff, F., 1919. "Dimension und äusseres Mass." *Math. Ann.*, *79*, 157–79.

Heine, H., 1870. "Ueber trigonometrische Reihen." *Crelle, Jl. Math.*, *71*, 353–65.

Hermite, C., 1883. "Sur quelques points de la theorie des fonctions." *Helsingfors, Acta, 12*, 69–94.

Hermite, C., 1891. *Cours de M. Hermite,* 4th ed. Paris, A. Hermann.

Hermite, C., 1916. "Briefe von Ch. Hermite an P. du Bois-Reymond aus den Jahren 1875–1888." *Arch. Math. Phys.*, (3) *24*, 193–220, 289–310.

Hermite, C., and T.-J. Stieltjes, 1905. *Correspondance d'Hermite et de Stieltjes.* 2 vols. Edited by B. Baillaud and H. Bourget. Paris, Gauthier-Villars.

Hilbert, D., 1891. "Ueber die stetige Abbildung einer Linie auf ein Flächenstück." *Math. Ann.*, *38*, 459–60.

Hobson, E. W., 1906. "On Absolutely Convergent Improper Double Integrals." *London Math. Soc. Proc.*, (2) *4*, 136–59.

Hobson, E. W., 1907. "On Repeated Integrals." *London Math. Soc. Proc.*, (2) *5*, 325–34.

Hobson, E. W., 1910. "On Some Fundamental Properties of Lebesgue Integrals in a Two-Dimensional Domain." *London Math. Soc. Proc.*, (2) *8*, 22–39.

Hobson, E. W., 1926–27. *The Theory of Functions of a Real Variable and the Theory of Fourier Series*, 2nd ed. 2 vols. Cambridge, The University Press.

Hölder, O., 1884. "Zur Theorie der trigonometrischen Reihen." *Math. Ann.*, *24*, 181–216.

Hoüel, J., 1870. Review of [Hankel 1870]. *Bull. Sci. Math.*, *1*, 117–24.

Hurwitz, A., 1903. "Über die Fourierschen Konstanten integrirbarer Funktionen." *Math. Ann.*, *57*, 425–46.

Jordan, C., 1881. "Sur la série de Fourier." *Paris, Ac. Sci. C.R.*, *92*, 228–30.

Jordan, C., 1882–87. *Cours d'analyse de l'École Polytechnique*, 1st ed. 3 vols. Paris, Gauthier-Villars.

Jordan, C., 1892. "Remarques sur les intégrales défines." *Liouville, Jl. Math.*, *(4) 8*, 69–99.

Jordan, C., 1893–96. *Cours d'analyse de l'École Polytechnique*, 2nd ed. 3 vols. Paris, Gauthier-Villars.

Jourdain, P. E. B., 1906. "The Development of the Theory of Transfinite Numbers," Pt 1. *Arch. Math. Phys.*, *(3) 10*, 254–81.

Jourdain, P. E. B., 1909. "The Development of the Theory of Transfinite Numbers," Pt 2. *Arch. Math. Phys.*, *14*, 289–311.

Jourdain, P. E. B., 1910. "The Development of the Theory of Transfinite Numbers," Pt 3. *Arch. Math. Phys.*, *16*, 21–43.

Jourdain, P. E. B., 1913a. "The Origin of Cauchy's Conceptions of a Definite Integral and of the Continuity of a Function." *Isis, 1*, 661–703.

Jourdain, P. E. B., 1913b. "The Development of the Theory of Transfinite Numbers," Pt 4. *Arch. Math. Phys.*, *22*, 1–21.

Königsberger, L., 1874. *Vorlesungen über die Theorie der elliptischen Functionen, nebst einer Einleitung in die allgemeine Functionenlehre.* Leipzig.

Köpcke, H. A., 1887. "Ueber Differentiirbarkeit und Anschaulichkeit der stetigen Functionen." *Math. Ann.*, *29*, 123–40.

Köpcke, H. A., 1889. "Ueber eine durchaus differentiirbare, stetige Function mit Oscillationen in jedem Intervalle." *Math. Ann.*, *34*, 161–71.

Köpcke, H. A., 1890. "Nachtrag zu dem Aufsatze 'Ueber eine durchaus differentiirbare, stetige Function . . .'." *Math. Ann.*, *35*, 104–09.

Kronecker, L., 1879. " . . . Notiz über Potenzreihen. . . . " *Berlin Ak. Monatsber.*, 53–58.

Kronecker, L., 1889. "Paul du Bois-Reymond." *Crelle, Jl. Math.*, *104*.

Lacroix, S. F., 1810–19. *Traité de calcul différentiel et de calcul intégral*, 2nd ed. 3 vols. Paris, Courcier.

Lagrange, J. L., 1772. "Sur une nouvelle espèce de calcul relatif à la différentiation et à l'intégration des quantités variables." *Nouveaux Mémoires de l'Académie royale des Sciences et Belles-Lettres de Berlin*, 185–221.

Lagrange, J. L., 1797. *Théorie des fonctions analytiques*. . . . Paris.

Lagrange, J. L., 1803. "Leçons sur le calcul des fonctions." *Paris, École Polyt. Jl.*, (1) *12*.

Lamarle, A. H. E., 1855. "Étude approfondie sur les deux équations fondamentales lim $(f(x+h) - f(x))/h = f'(x)$ et $dy = f'(x) \Delta x$." *Brux., Ac. Mem.*, *29*.

Lebesgue, H., 1898. "Sur l'approximation des fonctions." *Bull. Sci. Math.*, *22*, 278–87.

Lebesgue, H., 1899a. "Sur les fonctions de plusieurs variables." *Paris, Ac. Sci. C.R.*, *128*, 811–13.

Lebesgue, H., 1899b. "Sur quelques surfaces non réglées applicables sur le plan." *Paris, Ac. Sci. C.R.*, *128*, 1502–05.

Lebesgue, H., 1899c. "Sur la définition de l'aire d'une surface." *Paris, Ac. Sci. C.R.*, *129*, 870–73.

Lebesgue, H., 1900a. "Sur la définition de certaines intégrales de surface." *Paris, Ac. Sci. C.R.*, *131*, 867–70.

Lebesgue, H., 1900b. "Sur le minimum de certaines intégrales." *Paris, Ac. Sci. C.R.*, *131*, 935–37.

Lebesgue, H., 1901. "Sur une généralisation de l'intégrale définie." *Paris, Ac. Sci. C.R.*, *132*, 1025–28.

Lebesgue, H., 1902a. "Intégrale, longueur, aire." *Ann. Mat.*, (3) *7*, 231–359.

Lebesgue, H., 1902b. "Un théorème sur les séries trigonométriques." *Paris, Ac. Sci. C.R.*, *134*, 585–87.

Lebesgue, H., 1903a. "Sur les séries trigonométriques." *Paris, École Norm. Ann.*, (3) *20*, 453–85.

Lebesgue, H., 1903b. "Sur l'existence des dérivées." *Paris, Ac. Sci. C.R.*, *136*, 659–61.

Lebesgue, H., 1903c. "Sur une propriété des fonctions." *Paris, Ac. Sci. C.R.*, *137*, 1228–30.

Lebesgue, H., 1904. *Leçons sur l'intégration et la recherche des fonctions primitives*. Paris, Gauthier-Villars.

Lebesgue, H., 1905a. "Sur une condition de convergence des séries de Fourier." *Paris, Ac. Sci. C.R.*, *140*, 1378–81.

Lebesgue, H., 1905b. "Recherches sur la convergence des séries de Fourier." *Math. Ann.*, *61*, 251–80.

Lebesgue, H., 1906a. *Leçons sur les séries trigonométriques*. Paris, Gauthier-Villars.

Lebesgue, H., 1906b. "Sur les fonctions dérivées." *Roma, R. Acc. Lincei Rend.*, (5) 15_2, 3–8.

Lebesgue, H., 1907a. "Encore une observation sur les fonctions dérivées." *Roma, R. Acc. Lincei Rend.*, (5) *16₁*, 92–100.

Lebesgue, H., 1907b. "Sur la recherche des fonctions primitives par l'intégration." *Roma, R. Acc. Lincei Rend.*, (5) *16₁*, 283–90.

Lebesgue, H., 1907c. "Contribution a l'étude des correspondences de M. Zermelo." *Bull. Soc. Math. de France, 35*, 202–21.

Lebesgue, H., 1908. "Sur la méthode de M. Goursat pour la résolution de l'équation de Fredholm." *Paris, Soc. Math. Bull., 36*, 3–19.

Lebesgue, H., 1910a. "Sur l'intégrale de Stieltjes et sur les opérations linéaires." *Paris, Ac. Sci. C.R., 150*, 86–88.

Lebesgue, H., 1910b. "Sur l'intégration des fonctions discontinues." *Paris, École Norm. Ann.*, (3) *27*, 361–450.

Lebesgue, H., 1922. *Notice sur les travaux scientifiques de M. Henri Lebesgue.* Toulouse.

Lebesgue, H., 1926. "Notice sur la vie et les travaux de Camille Jordan." *Paris, Ac. Sci. Mém.*, (2) *58*, xxxix–lxvi.

Lebesgue, H., 1927. "Sur le développement de la notion d'intégrale." *Revue de métaphysique et de morale, 34*, 149–67.

Lebesgue, H., 1932. "Notice sur René-Louis Baire, Correspondant pour la Section de Géometrie." *Paris. Ac. Sci. C.R., 195*, 86–88.

Lebesgue, H., 1966. *Measure and the Integral.* Edited, with a biographical essay, by Kenneth O. May. San Francisco, Holden-Day.

Levi, B., 1906a. "Richerche sulle funzioni derivate." *Roma, R. Acc. Lincei Rend.*, (5) *15₁*, 433–38, 551–58.

Levi, B., 1906b. "Richerche sopra la funzioni derivate." *Roma, R. Acc. Lincei Rend.*, (5) *15₁*, 674–84.

Levi, B., 1906c. "Ancora alcune osservazione sulle funzioni derivate." *Roma, R. Acc. Lincei Rend.*, (5) *15₂*, 358–68.

Levi, B., 1906d. "Sopra l'integrazione delle serie." *Milano Ist. Lomb. Rend.*, (2) *39*, 775–80.

Levi, B., 1906e. "Sul principio di Dirichlet." *Palermo, Circ. Mat. Rend., 22*, 293–359.

Libri, G., 1831–33. "Mémoire sur les fonctions discontinues." *Crelle, Jl. Math., 7* (1831), 224–33; *10* (1833), 303–16.

Libri, G., 1842. "Mémoire sur l'emploi des fonctions discontinues dans l'analyse, pour la recherche des formules générales." *Paris, Ac. Sci. C.R., 15*, 401–11.

Lipschitz, R., 1864. "De explicatione per series trigonometricas instituenda functionum unius variabilis arbitrariarum, et praecipue earum, quae per variabilis spatium finitum valorum maximorum et minimorum numerum habent infinitum, disquisitio." *Crelle, Jl. Math., 63*, 296–308. Translated as [Montel 1912a].

Lipschitz, R., 1877–80. *Lehrbuch der Analysis.* 2 vols. Bonn, M. Cohen.

MacFarlane, Alexander, 1916. *Lectures on Ten British Mathematicians of the Nineteenth Century*, 1st ed. New York, John Wiley and Sons.

Manheim, J. H., 1964. *The Genesis of Point Set Topology*. New York, Macmillan.

May, Kenneth O., 1966. "Biographical Sketch of Henry Lebesgue." [Lebesgue 1966: 1–7].

Meschowski, H., 1961. *Denkweisen grosser Mathematiker: Ein Weg zur Geschichte der Mathematik*. Brunswick, Germany, F. Vieweg.

Meyer, Gustav Ferdinand, 1871. *Vorlesungen über die Theorie der bestimmten Integrale zwischen reellen Grenzen, mit vorzüglicher Berücksichtigung der von P. Gustav Lejeune-Dirichlet im Sommer 1858 gehaltenen Vorträge über bestimmte Integrale*. Leipzig, B. G. Teubner.

Mittag-Leffler, G., 1912. "Zur Biographie von Weierstrass." *Acta Math.*, *35*, 29–65.

Mittag-Leffler, G., 1923. "Weierstrass et Sonja Kowalewsky." *Acta Math.*, *39*, 133–98.

Montel, P., 1912a. "Recherches sur le développement en séries trigonométriques des fonctions arbitraires d'une variable et principalement de celles qui, dans un intervalle fini, admettent une infinité de maxima et de minima." *Acta Math.*, *36*, 281–95. Translation of [Lipschitz 1864].

Montel, P., 1912b. "Intégration et dérivation." *Encyclopédie des sciences mathématiques pures at appliquées*, Tome II, Vol. 1, Fasc. 2, pp. 171–209. Paris, Gauthier-Villars, and Leipzig, B. G. Teubner.

Moore, E. H., 1901a. "Concerning Harnack's Theory of Improper Definite Integrals." *Amer. Math. Soc. Trans.*, *2*, 296–330.

Moore, E. H., 1901b. "On the Theory of Improper Definite Integrals." *Amer. Math. Soc. Trans.*, *2*, 459–75.

Murphy, R., 1833. "On the Inverse Method of Definite Integrals with Physical Applications." *Cambridge, Phil. Soc. Trans.*, *4*, 353–408.

Osgood, W. F., 1896. "Ueber die ungleichmässige Convergenz und die gliedweise Integration der Reihen." *Göttingen Nachr.*, 288–91.

Osgood, W. F., 1897. "Non Uniform Convergence and the Integration of Series Term by Term." *Amer. Jl. Math.*, *19*, 155–90.

Pasch, M., 1887. "Ueber einzige Punkte der Functionentheorie." *Math. Ann.*, *30*, 132–54.

Peacock, George, 1833. "Report on the Recent Progress and Present State of Certain Branches of Analysis." *Brit. Ass. Rep.*, 185–352.

Peano, G., 1883. "Sulla integrabilità delle funzioni." *Torino Acc. Sci. Atti*, *18*, 439–46.

Peano, G., 1887. *Applicazione geometriche del calcolo infinitesimale*. Torino, Bocca.

Peano, G., 1890. "Sur une courbe qui remplit une aire." *Math. Ann.*, *36*, 157–60.

Pereno, Italo, 1897. "Sulle funzioni derivabili in ogni punto ed infinitamente oscillanti in ogni intervallo." *Giorn. Mat.*, *35*, 132–49.

Picard, Émile, 1891. *Traité d'analyse*, Vol. 1. Paris, Gauthier-Villars.

Piola, G., 1828. "Sulla teoria delle funzioni discontinue." *Modena Mem. Soc. Ital.*, *20*, 573–639.

Poincaré, H., 1883. "Sur les fonctions à espaces lacunaires." *Helsingfors, Acta*, *12*, 343–50.

Poincaré, H., 1892. "Sur les fonctions à espaces lacunaires." *Amer. Jl. Math.*, *14*, 201–21.

Pompeiu, D., 1907. "Sur les fonctions dérivées." *Math. Ann.*, *63*, 326–32.

Pringsheim, A., 1892. "Zur Theorie der Taylor'sche Reihe und der analytischen Functionen mit beschränktem Existenzbereich." *München, Ak. Sber.*, *22*, 211–45. Also in *Math. Ann.*, *42* (1893), 153–84.

Pringsheim, A., 1900. "Zur Theorie des Doppel-Integrals, des Green'-schen und Cauchy'schen Integralsatzes." *München, Ak. Sber.*, *29*, 39–62.

Pringsheim, A., and J. Molk, 1909. "Principes fondamentaux de la théorie des fonctions." *Encyclopédie des sciences mathématiques pures et appliquees*, Tome II, Vol. 1, Fasc. 1. Paris, Gauthier-Villars.

Raabe, J. L., 1839–47. *Die Differential- und Integralrechnung*. 2 vols. Zürich, Orell, Füssli und Cie.

Radon, J., 1913. "Theorie und Anwendungen der absolut additiven Mengenfunctionen." *Wien, Ak. Sber.*, *122*IIa, 1295–1438.

Rawson, R., 1848. "A New Mode of Representing Discontinuous Functions." *Manchester Lit. Phil. Soc. Mem.*, (2) *8*, 235–64.

Reiff, R., 1889. *Geschichte der unendlichen Reihen*. Tübingen, H. Laupp.

Riemann, Bernhard, 1851. "Grundlagen für eine allgemeine Theorie der Functionen einer veränderlichen complexen Grösse." Inaugural dissertation, Göttingen. Page references are from *Werke*, pp. 1–48.

Riemann, Bernhard, 1902. *Gesammelte Mathematische Werke und Wissenschaftlicher Nachlass*, 2nd ed., and *Nachträge*. Leipzig, B. G. Teubner.

Riesz, F., 1906. "Sur les ensembles de fonctions." *Paris, Ac. Sci. C.R.*, *143*, 738–41.

Riesz, F., 1907a. "Sur les systèmes orthogonaux de fonctions." *Paris, Ac. Sci. C.R.*, *144*, 615–19.

Riesz, F., 1907b. "Über orthogonale Funktionensysteme." *Göttingen Nachr.*, 116–22.

Riesz, F., 1909a. "Sur les suites de fonctions mesurables." *Paris, Ac. Sci. C.R.*, *148*, 1303–05.

Riesz, F., 1909b. "Sur les opérations fonctionnelles linéaires." *Paris, Ac. Sci. C.R.*, *149*, 974–77.

Riesz, F., 1910. "Untersuchungen über Systeme integrirbarer Funktionen." *Math. Ann.*, *69*, 449–97.

Riesz, F., 1949. "L'Évolution de la notion d'intégrale depuis Lebesgue." *Ann. Inst. Fourier*, *1*, 29–42.

Rubini, R., 1868. *Elementi di calcolo infinitesimale*, Vol. 1. Naples.

Scheeffer, L., 1884a. "Allgemeine Untersuchungen über Rectification der Curven." *Acta Math.*, *5*, 49–82.

Scheeffer, L., 1884b. "Zur Theorie der stetigen Funktionen einer reellen Veränderlichen." *Acta Math.*, *5*, 183–94.

Schlesinger, L., 1912. "Über Gauss' Arbeiten zur Funktionentheorie." *Materialen für eine wissenschaftliche Biographie von Gauss*, Pt 3. Leipzig, B. G. Teubner.

Schmidt, E., 1906. "Sur la puissance des systèmes orthogonaux." *Paris, Ac. Sci. C.R.*, *143*, 955–57.

Schoenflies, Arthur, 1900. "Die Entwickelung der Lehre von den Punktmannigfaltigkeiten." *Deutsch. Math. Ver. Jber.*, *8*.

Schoenflies, Arthur, 1901. "Ueber die oscillirenden differenzirbaren Functionen." *Math. Ann.*, *54*, 553–63.

Schwarz, H. A., 1870. Letter to Georg Cantor dated 25 February. [Meschkowski, 1961: 78–79].

Schwarz, H. A., 1872. "Zur Integration der partiellen Differentialgleichung $\partial^2 u/\partial x^2 + \partial^2 u/\partial y^2 = 0$. *Crelle, Jl. Math.*, *74*, 218–53.

Schwarz, H. A., 1873. "Neues Beispiel einer stetigen nicht differentiirbaren Function." *Schweiz. Natf. Ges. Verh.*, *56*, 252–58.

Seidel, P. L., 1850. "Note über eine Eigenschaft der Reihen, welche discontinuirliche Functionen darstellen." *München, Ak. Abh.*, *5*, 379–93.

Serret, J. A., 1868. *Cours de calcul différentiel et intégral*. 2 vols. Paris, Gauthier-Villars. Translated into German, with supplementation, as [Harnack 1884–85].

Singh, A. N., 1953. "The Theory and Construction of Non-differentiable Functions." *Squaring the Circle and Other Monographs*. New York, Chelsea.

Smith, H. J. S., 1875. "On the Integration of Discontinuous Functions." *London Math. Soc. Proc.*, *6*, 140–53.

Stieltjes, T.-J., 1887. "Exemple d'une fonction qui n'existe qu'a l'intérieur d'un cercle." *Bull. Sci. Math.*, (2) *11*, 46–51.

Stieltjes, T.-J., 1894. "Recherches sur les fractions continues." *Toulouse Ann.*, *8*, J.1–J.122.

Stokes, G. G., 1848. "On the Critical Values of the Sums of Periodic

Series." *Cambridge, Phil. Soc. Trans.*, 8_5, 533–83. Also in [Stokes 1880–1905: I, 236–313].

Stokes, G. G., 1880–1905. *Mathematical and Physical Papers.* 5 vols. Cambridge, At the University Press.

Stolz, O., 1881. "B. Bolzano's Bedeutung in der Geschichte der Infinitesimalrechnung." *Math. Ann.*, *18*, 256–79.

Stolz, O., 1884. "Ueber einen zu einer unendlichen Punktmenge gehörigen Grenzwerth." *Math. Ann.*, *23*, 152–56.

Stolz, O., 1893–99. *Grundzüge der Differential- und Integralrechnung.* 3 vols. Leipzig, B. G. Teubner.

Stolz, O., 1898. "Zur Erklärung der absolut convergenten uneigentlichen Integrale." *Wien, Ak. Sber.*, *107*IIa, 207–24.

Stolz, O., 1902a. "Zur Erklärung der Bogenlänge und des Inhaltes einer krummen Fläche." *Amer. Math. Soc. Trans.*, *3*, 23–33.

Stolz, O., 1902b. "Nachtrag zum Artikel: 'Zur Erklärung der Bogenlänge usw.' " *Amer. Math. Soc. Trans.*, *3*, 302–04.

Study, E., 1896. "Ueber eine besondere Classe von Funktionen einer reellen Veränderlichen." *Math. Ann.*, *47*, 298–316.

Tannery, Jules, 1887. Review of [Peano 1887]. *Bull. Sci. Math.*, *(2) 11*, 237–39.

Thomae, K. J., 1875. *Einleitung in die Theorie der bestimmten Integrale.* Halle.

Thomae, K. J., 1876. "Zur Definition des bestimmtes Integrales durch den Grenzwerth einer Summe." *Zeitschr. Math. Phys.*, *21*, 224–27.

Thomae, K. J., 1878. "Ueber bestimmte Integrale." *Zeitschr. Math. Phys.*, *23*, 67–68.

Van Vleck, E. B., 1908. "On Non-measurable Sets of Points with an Example." *Amer. Math. Soc. Trans.*, *9*, 237–44.

Veltmann, W., 1882. "Ueber die Anordnung unendlich vieler Singularitäten einer Function." *Zeitschr. Math. Phys.*, *27*, 176–79.

Vitali, G., 1903. "Sulla condizione di integrabilità delle funzioni." *Catania Acc. Gioen. Bull.*, *79*, 27–30.

Vitali, G., 1904a. "Sulla integrabilità delle funzioni." *Milano, Ist. Lomb. Rend.*, *(2) 37*, 69–73.

Vitali, G., 1904b. "Sui gruppi di punti." *Palermo, Circ. Mat. Rend.*, *18*, 116–26.

Vitali, G., 1904–05. "Sulle funzioni integrali." *Torino Acc. Sci. Atti, 40*, 1021–34.

Vitali, G., 1905a. "Sopra l'integrazione di serie di funzioni di una variabile reale." *Catania Acc. Gioen. Bull.*, *86*, 3–9.

Vitali, G., 1905b. *Sul problema della misura dei gruppi de punti di una retta.* Bologna.

Vitali, G., 1907. "Sull'integrazione per serie." *Palermo, Circ. Mat. Rend.*, *23*, 137–55.

Vitali, G., 1907–08. "Sui gruppi di punti e sulle funzioni di variabili reali." *Torino Acc. Sci. Atti.*, *43*, 229–46.

Volterra, V., 1881a. "Alcune osservazioni sulle funzioni punteggiate discontinue." *Giorn. Mat.*, *19*, 76–86.

Volterra, V., 1881b. "Sui principii del calcolo integrale." *Giorn. Mat.*, *19*, 333–72.

Volterra, V., 1889. "Sur une généralisation de la théorie des fonctions d'une variable imaginaire." *Acta Math.*, *12*, 233–86.

Weierstrass, Karl, 1880. "Zur Functionenlehre." *Berlin Ak. Monatsber*, 719–43.

Weierstrass, Karl, 1894–1927. *Mathematische Werke.* 7 vols. Berlin, Mayer and Müller.

Weierstrass, Karl, 1923a. "Briefe von K. Weierstrass an Paul du Bois-Reymond" [with notes by G. Mittag-Leffler]. *Acta. Math.*, *39*, 199–225.

Weierstrass, Karl, 1923b. "Briefe von K. Weierstrass an L. Koenigsberger." *Acta. Math.*, *39*, 226–39.

Weyl, H., 1909. "Über die Konvergenz von Reihen, die nach Orthogonalfunktionen fortschreiten." *Math. Ann.*, *67*, 225–45.

Young, G. C., 1916. "On Infinite Derivatives: An Essay." *Quart. Jl. Math.*, *47*, 127–75.

Young, W. H., 1903a. "Sets of Intervals on the Straight Line." *London Math. Soc. Proc.*, *35*, 245–68.

Young, W. H., 1903b. "On Closed Sets of Points as the Limit of a Sequence of Closed Sets of Points." *London Math. Soc. Proc.*, *35*, 269–82.

Young, W. H., 1903c. "A Note on Unclosed Sets of Points as the Limit of a Sequence of Closed Sets of Points." *London Math. Soc. Proc.*, *35*, 283–84.

Young, W. H., 1903d. "Overlapping Intervals." *London Math. Soc. Proc.*, *35*, 384–88.

Young, W. H., 1904a. "A Note on the Condition of Integrability of a Function of One Variable." *Quart. Jl. Math.*, *35*, 189–92.

Young, W. H., 1904b. "Open Sets and the Theory of Content." *London Math. Soc. Proc.*, (*2*) *2*, 16–51.

Young, W. H., 1904c. "The General Theory of Integration." *Roy. Soc. Proc.*, *73*, 445–49. Abstract of [Young 1905].

Young, W. H., 1904d. "On Upper and Lower Integration." *London Math. Soc. Proc.*, (*2*) *2*, 52–66.

Young, W. H., 1905. "On the General Theory of Integration." *Phil. Trans.*, *204*, 221–52. (The abstract of this is [Young 1904c].)

Index

Abel, N. H., 21–22, 165
Absolute continuity, 77–78, 118, 142–48, 183–84, 186, 189, 193
Additivity of set functions, 62, 66, 70, 73, 88, 88n, 94, 103, 107, 121–23, 147, 150, 184–87
Ampère, A. M., 43–44, 48, 49
Appell, P., 97, 99
Arzelà, C., 91–93, 96, 128, 147, 149, 177
Arzelà's Lemma, 116, 149
Ascoli, G., 25, 41, 85n

Baire, R., 83n, 108n, 117–18, 120, 127, 137, 137n
Baire function, 127
Bendixson, I., 72–73, 104
Bernoulli, D., 4–6
Bertrand, J., 44n
Bettazzi, R., 91
Bolzano, B., 16, 45n, 81
Bonnet, O., 27, 48, 50n
Borel, E.
    work discussed, 97–108, 170n
    mentioned, 33, 64, 65, 70, 72, 73, 86, 96, 109n, 110, 120–24 *passim*, 132, 137, 146, 147, 148
Bounded variation, function of, 42, 48, 83–85, 140, 142, 172, 175, 177, 182, 186–88
Boussinesq, J., 90n
Brodén, T., 109–10
Burkill, J. C., 131

Cantor, G.
    work discussed, 22–25, 39n, 62–63, 71–74
    mentioned, 20, 27, 29, 33, 36, 37, 38, 47, 49, 64–70 *passim*, 76, 82, 86, 87, 94, 96, 104, 106, 109, 118, 147
Cantor ternary set, 72n
Carathéodory, C., 63n
Cauchy, A. L.
    work discussed, 9–12, 21–22, 44, 91
    mentioned, 3, 13, 15, 29, 32, 50, 67, 70, 88, 189
Cellérier, C., 45n
Cesàro, E., 165
Chain of intervals, 134–35, 140, 146
Closed set, 72
Condensation of singularities, 45, 82–83
Convergence
    in measure, 112, 116, 177–78
    mean, 176, 177–78
Curve, space-filling, 91, 189–91
Curve length, as integral, 79–85, 130, 140–41

d'Alembert, J., 4–5
Daniell, P. J., 132n
Darboux, G., 27–28, 41, 46, 47, 50, 52, 93, 152, 153
Dedekind, R., 20
de la Vallée-Poussin, C.
    work discussed, 110, 142, 151n, 154–57, 161–62, 161n

mentioned, 144, 158, 159, 167n
Denjoy, A., 137, 137n
Derived set, 15, 71–72
Dini, U.
    work discussed, 25, 36–37, 47–54
    mentioned, 42, 55–60 *passim*, 71,
        78, 79, 128, 131
Dirichlet, P. G. Lejeune-
    work discussed, 11–17, 90n
    mentioned, 24, 29, 32, 34, 67, 68,
        80, 83, 84, 118, 184
Dirichlet's nonintegrable function, 16,
    30–31, 118, 127, 153, 156
Dirksen, E. H., 81n
Discrete set, 59
Dominated Convergence Theorem,
    118n
Drach, J., 103, 121, 132
du Bois-Reymond, P.
    work discussed, 25–26, 33–36, 41–
        42, 51, 58, 75–76, 80–84, 111–13
    mentioned, 28, 46, 53, 59, 63, 67–
        72 *passim*, 78, 92, 115, 131, 138,
        144, 164, 165, 168
Duhamel, J. M. C., 44n, 81–82

Euler, L., 3–4, 6, 8, 16, 29, 97

Fatou, P., 163, 168–72, 173, 174, 175,
    176
Fatou's Lemma, 170–71, 176
Fejér, L., 166–67
First category set, 108n
First species function, 48
First species set, 108, 108n
Fischer, E., 163, 176–78
Fourier, J.
    work discussed, 5–9
    mentioned, 11, 12, 16, 21, 22, 26, 29,
        53, 68, 76, 164
Fréchet, M., 173
Freycinet, C., 44n
Fubini, G., 154, 160–61, 161n
Fubini's Theorem, 89–93, 95–96, 154–
    62
Function. *See also* Absolute continuity
    of bounded variation, 42, 48, 83–85,
        140, 142, 172, 175, 177, 182, 186–88
    first species, 48
    of null integral, 51, 59, 59n
    with dense intervals of invariability,
        52, 55, 57, 74–75, 79, 118
    with nonintegrable derivative, 52–
        53, 57, 68, 93, 108–10
    Baire, 127
    semi-continuous, 151, 151n

harmonic, 165, 168–72 *passim*
Fundamental Theorem of Calculus
    I–III, 10
    I, 59, 131, 137, 166, 169
    II, 50–53, 57, 78–79, 108, 129–31,
        134, 136, 137
    III, 60, 74–75

Galois, E., 44n
Gauss, K. F., 16, 21, 26n
Gilbert, P., 45–46, 50, 90n
Goursat, E., 98, 121

Hadamard, J., 121, 182
Hahn, H., 137
Halphén, G., 61
Hankel, H.
    work discussed, 28–33, 44–45
    mentioned, 24, 27, 34–40 *passim*, 46,
        47, 50, 55, 59, 68, 83
Harmonic function, 165, 158–72 *passim*
Harnack, A.
    work discussed, 37, 58–61, 63–66,
        75–79, 90–93, 167–68
    mentioned, 42, 58, 72, 74, 83, 86, 96,
        106, 116, 118, 124, 130, 131, 136,
        138, 142, 144, 145, 176, 177, 178
Hausdorff, F., 63n
Heine, H., 22–25, 27, 28, 33, 101n, 116
Heine-Borel Theorem, 101n, 147, 150
Hermite, C., 90n, 93n, 97, 121–22
Hilbert, D., 172–74, 186, 189–90, 191,
    192, 193
Hobson, E. W., 156, 158–60, 161n
Hölder, O., 75–77, 78, 137, 142, 144
Hurwitz, A., 167

Integrable set, 58–59
Integral
    as area, 8, 66, 70, 87, 95, 110, 124
    upper and lower, 40–41

Jordan, C.
    work discussed, 84–85, 90–96
    mentioned, 48, 62, 65n, 66, 67, 70,
        86, 97, 105, 106, 107, 121, 122, 126,
        130n, 146, 147, 152, 154, 155, 182

Königsberger, L., 46–47
Köpcke, H. A., 109, 109n, 110
Kowalewsky, S., 69
Kronecker, L., 111–12, 116, 117, 177

L^p-spaces, 173, 177–78
Lacroix, S. F., 44n
Lagrange, J. L., 5, 43, 97

Lamarle, A. H. E., 44, 50
Lebesgue, H.
  work discussed, 68, 96, 96n, 118,
    118n, 120–46, 157–58, 163–68,
    182–86, 191, 192n
  mentioned *passim*
Length of curve, as integral, 79–85,
  130, 140–41
Levi, B., 134, 136, 139, 160, 161, 171
Lipschitz, R., 8, 10, 14–15, 16, 24, 29,
  32, 90n
Lower integral, 40–41

Metric density, 138, 139n
Moore, E. H., 143–44
Müller, F., 40n

Nonintegrable derivative, 52–53, 57,
  68, 93, 108–10
Nowhere dense set, 13–15, 28–32, 37–
  40, 44, 53, 54, 55–56, 58–59, 71,
  73, 80–81
Null integrals, 51, 59, 59n

Osgood, W. F., 112–15, 117, 118, 128,
  148, 149
Osgood's Lemma, 115, 128, 149

Parseval's equality, 167–68, 171
Pasch, M., 59n
Peano, G.
  work discussed, 86–91
  mentioned, 41, 65n, 70, 93n, 94, 95,
    95n, 106, 107, 124, 185, 189, 191
Pereno, I., 109
Perfect set, 72–74
Picard, E., 91, 121
Poincaré, H., 97–98, 99, 100
Pompeiu, D., 109n
Pringsheim, A., 99, 158, 172

Raabe, J. L., 44n
Radon, J., 179, 185, 186–94
Radon-Nikodym Theorem, 189
Reducible set, 58–59, 72
Riemann, B.
  work discussed, 16–20
  mentioned *passim*
Riemann's integrable function, 18–19,
  29, 34, 45
Riesz, F., 163, 172–78, 182, 187, 191
Riesz-Fischer Theorem, 174–76
Rubini, R., 44n

Scheeffer, L., 49n, 74–75, 82–83, 85,
  118, 141, 145

Schmidt, E., 173
Schoenflies, A., 86, 106–08, 110, 119
Schwarz, H. A., 25, 26, 168
Second category set, 108, 108n
Seidel, P., 22
Semi-continuous function, 151, 151n
Serret, J. A., 44n
Set
  nowhere dense, 13–15, 28–32, 37–
    40, 44, 53, 54, 55–56, 58–59, 71,
    73, 80–81
  derived, 15, 71–72
  first species, 25, 28–29, 34–39, 54,
    58, 71, 75–76, 78, 80–82
  integrable, 58–59
  reducible, 58–59, 72
  discrete, 59
  closed, 72
  Cantor ternary, 72n
  perfect, 72–74
  first category, 108n
  second category, 108, 108n
Smith, H. J. S., 37–41, 56, 58
Space-filling curve, 91, 189–91
Stieltjes, T.-J., 98, 179–82, 183, 186
Stokes, G., 22
Stolz, O., 61–62, 61n, 68, 81, 85, 86,
  93n, 106, 142, 156, 159, 160
Study, E., 85

Tannery, J., 93n, 103
Term-by-term integration, 7, 21–23,
  25, 27–28, 42, 110–18, 118n, 148,
  161, 164–65, 175
Thomae, K. J., 40, 52, 90–91, 94, 158

Upper integral, 40–41

Van Vleck, E. B., 123n
Veltmann, W., 58
Vitali, G.
  work discussed, 145–48, 183–84
  mentioned, 40, 59, 123n, 142, 150,
    177
Volterra, V.
  work discussed, 56–58, 184–85
  mentioned, 41, 54, 58, 68, 70, 73,
    79, 93, 108, 109, 128, 145

Weierstrass, K.
  work discussed, 22, 45–47, 67–70
  mentioned, 82, 86, 97, 99, 109, 111
Weyl, H., 177

Young, G. C., 148
Young, W. H., 40, 59, 115, 127, 148–54

CHELSEA

SCIENTIFIC

BOOKS

THEORIE DES OPERATIONS LINEAIRES
By S. BANACH
—1933-63. xii + 250 pp. 5⅜x8. 8284-0110-1.

DIFFERENTIAL EQUATIONS
By H. BATEMAN

CHAPTER HEADINGS: I. Differential Equations and their Solutions. II. Integrating Factors. III. Transformations. IV. Geometrical Applications. V. Diff. Eqs. with Particular Solutions of a Specified Type. VI. Partial Diff. Eqs. VII. Total Diff. Eqs. VIII. Partial Diff. Eqs. of the Second Order. IX. Integration in Series. X. The Solution of Linear Diff. Eqs. by Means of Definite Integrals. XI. The Mechanical Integration of Diff. Eqs.

—1917-67. xi + 306 pp. 5⅜x8. 8284-0190-X.

MEASURE AND INTEGRATION
By S. K. BERBERIAN

A highly flexible graduate level text. Part I is designed for a one-semester introductory course; the book as a whole is suitable for a full-year course. Numerous exercises.

Partial Contents: PART ONE: I. Measures. II. Measurable Functions. III. Sequences of Measurable Functions. IV. Integrable Functions. V. Convergence Theorems. VI. Product Measures. VII. Finite Signed Measures. PART TWO: VIII. Integration over Locally compact Spaces (. . . The Riesz-Markoff Representation Theorem, . . .). IX. Integration over Locally Compact Groups (Topological Groups, . . . , Haar Integral, Convolution, The Group Algebra, . . .). BIBLIOGRAPHY. INDEX.

—1965-70. xx + 312 pp. 6x9. 8284-0241-8.

L'APPROXIMATION
By S. BERNSTEIN and CH. de LA VALLÉE POUSSIN

TWO VOLUMES IN ONE:

Leçons sur les Propriétés Extrémales et la Meilleure Approximation des Fonctions Analytiques d'une Variable Réelle, *by Bernstein.*

Leçons sur l'approximation des Fonctions d'une Variable Réelle, *by Vallée Poussin.*

—1925/19-69. 363 pp. 6x9. 8284-0198-5. 2 v. in 1.

LECTURES ON THE CALCULUS OF VARIATIONS
By O. BOLZA

A standard text by a major contributor to the theory. Suitable for individual study by anyone with a good background in the Calculus and the elements of Real Variables.

—2nd (corr.) ed. 1961. 280 pp. 5⅜x8. 8284-0145-4.

VORLESUNGEN UEBER VARIATIONSRECHNUNG
By O. BOLZA

A standard text and reference work, by one of the major contributors to the theory.

—1909-63. ix + 715 pp. 5⅜x8.   8284-0160-8.

THEORIE DER KONVEXEN KOERPER
By T. BONNESEN and W. FENCHEL

"Remarkable monograph."
         —*J. D. Tamarkin, Bulletin of the A. M. S.*
—1934. 171 pp. 5½x8½.   8284-0054-7.

ALGEBRAIC THEORY OF MEASURE AND INTEGRATION
By C. CARATHÉODORY

Translated from the German by FRED E. J. LINTON. By generalizing the concept of point function to that of a function over a Boolean ring ("soma" function), Prof. Carathéodory gives an algebraic treatment of measure and integration.

—1963. 378 pp. 6x9.   8284-0161-6.

VORLESUNGEN UBER REELLE FUNKTIONEN
By C. CARATHÉODORY

This great classic is at once a book for the beginner, a reference work for the advanced scholar and a source of inspiration for the research worker.

—3rd ed. (c.r. of 2nd). 1968. 728 pp. 5⅜x8. 8284-0038-5.

FORMULAS AND THEOREMS IN PURE MATHEMATICS

By G. S. CARR

Second edition, with an introduction by Jacques Dutka.

Over 6,000 formulas and results, systematically arranged, with indications or outlines of proofs, and references to the original periodical literature. Elementary through advanced results are covered, including even some quite esoteric topics (e.g., linkages and link-works). There is a most extensive and detailed index.

Partial Contents: PART I. Mathematical Tables. II. Algebra. III. Theory of Equations. IV. Plane Trigonometry. V. Spherical Trigonometry. VI. Elementary Geometry. VIII. Differential Calculus. IX. Integral Calculus. X. Calculus of Variations. XI. Differential Equations. XII. Calculus of Finite Differences. XIII. Analytic Geometry of the Plane. XIV. Analytic Geometry of Space. INDEX. FOLD-OUT PLATES.

It was the first edition of this work that inspired Ramanujan: "Through the new world opened to him [by Carr's book] Ramanujan went raging with delight. It was this book that awakened his genius."—*The Collected Papers of S. Ramanujan.*

—2nd ed. 1970. xxxvi + 935 pp. 5⅜x8. 8284-0239-6.

COLLECTED PAPERS (OEUVRES)

By P. L. CHEBYSHEV

One of Russia's greatest mathematicians, Chebyshev (Tchebycheff) did work of the highest importance in the Theory of Probability, Number Theory, and other subjects. The present work contains his post-doctoral papers (sixty in number) and miscellaneous writings. The language is French, in which most of his work was originally published; those papers originally published in Russian are here presented in French translation.

—1962. Repr. of 1st ed. 1,480 pp. 5½x8¼. 8284-0157-8.

Two vol. set.

THEORIE DER CONGRUENZEN

By P. L. CHEBYSHEV

This work, subtitled *Elemente der Zahlentheorie*, is the only of Chebyshev's writings not included in his Oeuvres (see above).

—1889-1970. xvii + 313 + 31 pp. 5⅜x8.

THE DOCTRINE OF CHANCES
By A. DE MOIVRE

In the year 1716 Abraham de Moivre published his *Doctrine of Chances*, in which the subject of Mathematical Probability took several long strides forward. A few years later came his *Treatise of Annuities*. When the third (and final) edition of the *Doctrine* was published in 1756 it appeared in one volume together with a revised edition of the work on Annuities. It is this latter two-volumes-in-one that is here presented in an exact photographic reprint.

—3rd ed. 1756-1967. xi + 348 pp. 6x9. 8284-0200-0.

DE MORGAN. See D. E. SMITH

COLLECTED MATHEMATICAL PAPERS
By L. E. DICKSON

—1969. 4 vols. Approx. 3,400 pp. 6½x9¼.

HISTORY OF THE THEORY OF NUMBERS
By L. E. DICKSON

"A monumental work . . . Dickson always has in mind the needs of the investigator . . . The author has [often] expressed in a nut-shell the main results of a long and involved paper *in a much clearer way than the writer of the article did himself*. The ability to reduce complicated mathematical arguments to simple and elementary terms is highly developed in Dickson."—*Bulletin of A. M. S.*

—Vol. I (Divisibility and Primality) xii + 486 pp. Vol. II (Diophantine Analysis) xxv + 803 pp. Vol. III (Quadratic and Higher Forms) v + 313 pp. 5⅜x8. 8284-0086-5.

Three vol. set. **$22.50**

STUDIES IN THE THEORY OF NUMBERS
By L. E. DICKSON

—1930-62. viii + 230 pp. 5⅜x8. 8284-0151-9.

ALGEBRAIC NUMBERS
By L. E. DICKSON, et al.

TWO VOLUMES IN ONE.

LECTURES ON ERGODIC THEORY

By P. R. HALMOS

CONTENTS: Introduction. Recurrence. Mean Convergence. Pointwise Convergence. Ergodicity. Mixing. Measure Algebras. Discrete Spectrum. Automorphisms of Compact Groups. Generalized Proper Values. Weak Topology. Weak Approximation. Uniform Topology. Uniform Approximation. Category. Invariant Measures. Generalized Ergodic Theorems. Unsolved Problems.

"Written in the pleasant, relaxed, and clear style usually associated with the author. The material is organized very well and painlessly presented."
—Bulletin of the A.M.S.

—1956-60. viii + 101 pp. 5⅜x8.   8284-0142-X.

ELEMENTS OF QUATERNIONS

By W. R. HAMILTON

Sir William Rowan Hamilton's last major work, and the second of his two treatises on quaternions.

—3rd ed. 1899/1901-68. 1,185 pp. 6x9. 8284-0219-1.
Two vol. set.

RAMANUJAN:
Twelve Lectures on His Life and Works

By G. H. HARDY

The book is somewhat more than an account of the mathematical work and personality of Ramanujan; it is one of the very few full-length books of "shop talk" by an important mathematician.

—1940-59. viii + 236 pp. 6x9.   8284-0136-5.

GRUNDZUEGE DER MENGENLEHRE

By F. HAUSDORFF

The original, 1914 edition of this famous work contains many topics that had to be omitted from later editions, notably, the theories of content, measure, and discussion of the Lebesgue integral. Also, general topological spaces, Euclidean spaces, special methods applicable in the Euclidean plane, the algebra of sets, partially ordered sets, etc.

—1914-49. 484 pp. 5⅜x8.   8284-0061-X.

Grundzüge Einer Allgemeinen Theorie der
LINEAREN INTEGRALGLEICHUNGEN
By D. HILBERT
—1912-53. 306 pp. 5½x8¼.   8284-0091-1.

GEOMETRY AND THE IMAGINATION
By D. HILBERT and S. COHN-VOSSEN
Translated from the German by P. NEMENYI.

"A fascinating tour of the 20th century mathematical zoo.... Anyone who would like to see proof of the fact that a sphere with a hole can always be bent (no matter how small the hole), learn the theorems about Klein's bottle—a bottle with no edges, no inside, and no outside—and meet other strange creatures of modern geometry will be delighted with Hilbert and Cohn-Vossen's book."
—*Scientific American.*

"Should provided stimulus and inspiration to every student and teacher of geometry."—*Nature.*

"A mathematical classic.... The purpose is to make the reader *see* and *feel* the proofs.... readers can penetrate into higher mathematics with ... pleasure instead of the usual laborious study."
—*American Scientist.*

"Students, particularly, would benefit very much by reading this book ... they will experience the sensation of being taken into the friendly confidence of a great mathematician and being shown the real significance of things."—*Science Progress.*

"A person with a minimum of formal training can follow the reasoning.... an important [book]."
—*The Mathematics Teacher.*
—1952. 358 pp. 6x9.   8284-0087-3.

GESAMMELTE ABHANDLUNGEN
(Collected Papers)
By D. HILBERT

Volume I (Number Theory) contains Hilbert's papers on Number Theory, including his long paper on Algebraic Numbers. Volume II (Algebra, Invariant Theory, Geometry) covers not only the topics indicated in the sub-title but also papers on Diophantine Equations. Volume III carries the sub-title: Analysis, Foundation of Mathematics, Physics, and Miscellaneous Papers.
—1932/35-66. 1,457 pp. 6x9. 8284-0195-0.
Three vol. set

CHELSEA SCIENTIFIC BOOKS

THE MATHEMATICAL THEORY OF THE TOP,
by F. KLEIN. See SIERPINSKI

FAMOUS PROBLEMS, and other monographs
By KLEIN, SHEPPARD, MacMAHON, and MORDELL

FOUR VOLUMES IN ONE.

FAMOUS PROBLEMS OF ELEMENTARY GEOMETRY, by *Klein*. A fascinating little book. A simple, easily understandable, account of the famous problems of Geometry—The Duplication of the Cube, Trisection of the Angle, Squaring of the Circle—and the proofs that these cannot be solved by ruler and compass—presentable, say, before an undergraduate math club (no calculus required). Also, the modern problems about transcendental numbers, the existence of such numbers, and proofs of the transcendence of *e*.

FROM DETERMINANT TO TENSOR, by *Sheppard*. A novel and charming introduction. Written with the utmost simplicity. PT I. Origin of Determinants. II. Properties of Determinants. III. Solution of Simultaneous Equations. IV. Properties. V. Tensor Notation. PT II. VI. Sets. VII. Cogredience, etc. VIII. Examples from Statistics. IX. Tensors in Theory of Relativity.

INTRODUCTION TO COMBINATORY ANALYSIS, by *MacMahon*. A concise introduction to this field. Written as introduction to the author's two-volume work.

THREE LECTURES ON FERMAT'S LAST THEOREM, by *Mordell*. This famous problem is so easy that a high-school student might not unreasonably hope to solve it; it is so difficult that tens of thousands of amateur and professional mathematicians, Euler and Gauss among them, have failed to find a complete solution. Mordell's very small book begins with an impartial investigation of whether Fermat himself had a solution (as he said he did) and explains what has been accomplished. This is one of the masterpieces of mathematical exposition.

—2nd ed. 1962. 350 pp. 5⅜x8. Four vols. in one.

8284-0108-X.　Cloth
8284-0166-7.　Paper

VORLESUNGEN UEBER NICHT-EUKLIDISCHE GEOMETRIE

By F. KLEIN

—1928-59. xii + 326 pp. 5x8.　　8284-0129-2.

ANALYTIC THEORY OF CONTINUED FRACTIONS

By H. S. WALL

Partial Contents: CHAP. I. The c. f. as a Product of Linear Fractional Transformations. II. Convergence Theorems. IV. Positive-Definite c. f. VI. Stieltjes Type c. f. VIII. Value-Region Problem. IX. J-Fraction Expansions. X. Theory of Equations. XII. Matrix Theory of c. f. XIII. C. f. and Definite Integrals. XIV. Moment Problem. XVI. Hausdorff Summability. XIX. Stieltjes Summability. XX. The Padé Table.

—1948-67. xiv + 433 pp. 5⅜x8.   8284-0207-8.

LOGIC, COMPUTERS, AND SETS

By H. WANG

Partial Contents: GENERAL (EXPOSITORY) SKETCHES (*Chaps. I-VI*) : I. The Axiomatic Method (§ 2. The problem of adequacy, . . . , §6. Gödel's Theorems, . . .). II. Eighty Years of Foundational Studies. III. The Axiomatization of Arithmetic (§2. Grassmann's calculus, . . . , § 6. Dedekind and Frege). V. Computation (§ 1. Concept of computability, . . . , § 6. Control of errors in calculating machines). CALCULATING MACHINES (*Chaps. VI-X*) : VI. A Variant to Turing's Theory. VII. Universal Turing Machines. VIII. The Logic of Automata. IX. Toward Mechanical Mathematics. FORMAL NUMBER THEORY (*Chaps. XI-XV*) : XII. Many-Sorted Predicate Calculi. XIII. Arithmetization of Metamathematics (§ 1. Gödel numbering, . . . , § 4. Arithmetic translations of axiom systems). XIV. Ackermann's Consistency Proof. . . . IMPREDICATIVE SET THEORY (*Chaps. XVI-XX*) : XVI. Different Axiom Systems . . . XVIII. Truth Definitions and Consistency Proofs . . . PREDICATIVE SET THEORY (*Chaps. XXI-XXV*) : XXII. Undecidable Sentences. XXIII. Formalization of Mathematics. . . .

Originally published under the title: *A Survey of Mathematical Logic.*

—1962-71. x + 651 pp. 6x9. 8284-0245-0.

A HISTORY OF THE CALCULUS OF VARIATIONS IN THE 18th CENTURY
By R. WOODHOUSE

"Those interested in the beginnings of the Calculus of Variations should read the book by Lagrange and the—unfortunately, very rare—book by Woodhouse. Euler's masterpiece [*Methodus* ...] will then offer him very little difficulty. This work of Euler's is distinguished by the abundance of its examples and is, to this day, one of the most interesting books that the mathematical literature has to offer."—C. CARATHEODORY, in *Variationsrechnung und Partielle Differentialgleichungen erster Ordnung*.

—I. TODHUNTER

—1811-1965. ix + x + 154 pp. 5x8⅜. 8284-0177-2.

BEGINNER'S BOOK OF GEOMETRY
By G. C. YOUNG and W. H. YOUNG

An elementary beginner's book by two distinguished mathematicians. Unlike Birkhoff's *Basic Geometry*, which is intended for teenagers, this book is truly for young children.

"The right one from a psychological point of view ... Its main object is to awaken the pupil's mind to the ideas by which we classify the properties of space; this is done by directions in paperfolding, in dissection of areas, in the construction of solid models, and the like. At the same time, various theorems are stated and proved."—*Nature*.

—2nd ed. 1905-70. xvi + 222 pp. 4⅝x6⅜. 8284-0231-0.

THE THEORY OF SETS OF POINTS
By W. H. YOUNG and G. C. YOUNG

—1906-71. Appr. 402 pp. 5⅜x8.